Over 2,000 Places

to see the

Total Solar Eclipse
April 8, 2024

Eat, Explore Parks,

and

Visit Attractions

  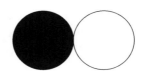

Craig Shields

**Over 2,000 Places to see the Total Solar Eclipse April 8, 2024
Eat, Explore Parks, and Visit Attractions**

A Clock Press Book copyright © 2024
by Craig Shields
published by Clock Press

ISBN: 978-0-9846718-6-1

Eclipse predictions courtesy of Fred Espenak,
NASA/Goddard Space Flight Center, from eclipse.gsfc.nasa.gov

Maps courtesy of the U.S. Geological Survey and NationalAtlas.gov

Cover image: Niagara Falls, NY

Disclaimer
Event times listed in this book are shown for reference and planning only. Venues may change schedules at any time and may not be open on eclipse day. Confirm availability in advance and observe local conditions for any activity. Locations and destinations listed in this book may not be suitable or safe for any activity. It is up to the reader to select an eclipse viewing site that is appropriate to their needs and safe for their planned activities. The author assumes no responsibility for the dangers, risks and liability that may result from the use of the information in this book.

It is not the purpose of this guide to reprint all the information that is otherwise available to amateur astronomers, eclipse chasers and the general public, but to complement, amplify and supplement other sources. This book does not attempt to cover all possible eclipse watching activities and events.

Every effort has been made to make this book as complete and as accurate as possible. However, there may be mistakes both typographical and in content. Therefore, this text should be used only as a general guide and in combination with other texts, atlases, maps and official information from property owners and governing authorities regarding eclipse watching activities.

The purpose of this book is to educate and entertain. The author and publisher shall have neither liability nor responsibility to any person or entity with respect to any loss or damage caused or alleged to be caused directly or indirectly by the information contained in this book.

The reader is encouraged to use common sense, obey all traffic laws, exercise patience and practice safe methods of viewing the total solar eclipse on April 8, 2024.

A Clock Press Book
www.clockpress.com

Table of Contents

Page

Introduction..5
Eye Safety..7
    How to know when it is safe to look without eye protection.....................8
    Diagram: Phases of the Total Solar Eclipse 9
Weather 11
Navigating the Path: Highways and Edge Cities 15
Texas - Oklahoma..19
    Places to Eat 20
    Parks and Attractions 45
Arkansas - Missouri - Kentucky...65
    Places to Eat 66
    Parks and Attractions 81
Illinois..95
    Places to Eat 96
    Parks and Attractions 103
Indiana..109
    Places to Eat 110
    Parks and Attractions 119
Ohio - Pennsylvania..131
    Places to Eat 132
    Parks and Attractions 143
New York - Vermont - New Hampshire - Maine................157
    Places to Eat 158
    Parks and Attractions 169
Resources 183
Glossary of Eclipse Terms 187

Introduction

On April 8, 2024, the United States will find itself under the path of a total solar eclipse, an extraordinary celestial event unseen since 2017. As the Moon's shadow arcs across the country, it will carve a path of totality through thirteen states, prominently featuring Texas, Oklahoma, Arkansas, Missouri, Kentucky, Illinois, Indiana, Ohio, Pennsylvania, New York, Vermont, New Hampshire, and Maine. Although the eclipse will briefly touch parts of Tennessee and Michigan, and extend beyond the U.S. into regions of Mexico and Canada, this guidebook focuses primarily on the aforementioned thirteen states. Within these areas, observers will be enthralled by a unique twilight as the Moon fully obscures the Sun, unveiling its stunning corona for nearly four and a half minutes. A partial eclipse will also be visible from most other parts of North America.

This event is particularly special because a similar total solar eclipse, spanning from coast to coast, won't happen again until August 12, 2045. With that in mind, this extraordinary spectacle affords millions of Americans a once-in-a-lifetime opportunity to witness a total solar eclipse within a manageable few hours' drive.

The book starts with a brief overview of eye safety as it relates to eclipse viewing. I've placed this information in the front to highlight the importance of safe eclipse viewing.

Next, the guide discusses weather patterns and how to identify the locations with the best prospects for clear skies on April 8. Weather is going to play a significant role in this eclipse, and since April is normally a month of very active weather in North America, it's important to know how to find clear sky on eclipse day.

Following that, mobility and cities on the edge of the eclipse path are addressed with a look at the highways you can use to stay mobile while remaining within the path of totality. Due to the nature of spring weather in North America, mobility will be a key factor for viewing this eclipse in the United States. In the case of cities near the edges of the boundary of the eclipse, the book lists places to eat, parks, and attractions that are all well within the path of totality in the edge cities. Please see the mobility section and the state sections for detailed information.

Then the directory sections begin, starting with Texas in the southwest and working northeast. Each state or multiple state section begins with a map showing the path of totality, a brief overview of the eclipse track, followed by the directory of the best places to eat, then the parks and attractions, organized by cities and listed alphabetically.

Eclipse start times, duration and end times are shown for each city and many parks and attractions. Where the park or attraction does not include the eclipse details, those details are identical or nearly identical to the city that the destination is listed under. Some parks and attractions are located within 30-50 miles of the listed city. If the duration of the eclipse at these locations is significantly different from that of the city, this information is included in the destination's description.

Ending the book are the resources section and a glossary of eclipse terms. The resources section lists all the National Weather Service offices and their web addresses in addition to many other websites specifically related to this eclipse, weather tools, online maps, and more.

Eye Safety

What is a Solar Filter?
A solar filter is a product designed and sold expressly for the purpose of protecting eyes and optical equipment from the harmful rays of the sun. A solar filter can take many forms. Sheets or rolls of film, coated glass mounted in "cells" for binoculars and telescopes, cardboard eclipse glasses, and solar viewers are a few of the types of solar filters available.

Different materials produce different colors and kinds of images of the sun. Eclipse glasses are usually black polymer and they give an orange color to the image. Other kinds of solar films and metal coated glass may give a white or blue appearance to the solar image. Welders glass shade #14 will give a green image of the sun.

Where to Get Solar Filters Online
Rainbow Symphony
rainbowsymphonystore.com

Seymour Solar
seymoursolar.com

Thousand Oaks Optical
thousandoaksoptical.com

Do not look at any of the partial phases of a solar eclipse without appropriate eye protection.

Don't forget, smartphone cameras can be damaged by the sun. Use a solar filter.

How to Use a Solar Filter
Before using a solar filter check it carefully for damage or defects. Read the instructions that came with the product and verify that this product is appropriate for your intended use. With any solar filter you select, the image of the solar disk should not be overly bright. If the image is uncomfortably bright immediately look away and try to determine what the problem is. Do not use a damaged or defective solar filter.

If you have ever burned leaves with a magnifying glass, you know about the focusing power of lenses.
DO NOT POINT OPTICAL DEVICES AT THE SUN WITHOUT A SOLAR FILTER.

Always put the Filter Closest to the Sun

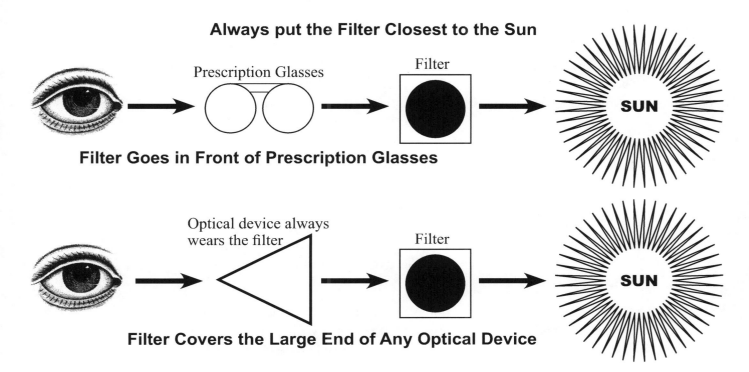

Prescription Glasses

Filter

SUN

Filter Goes in Front of Prescription Glasses

Optical device always wears the filter

Filter

SUN

Filter Covers the Large End of Any Optical Device

How do I know when it's OK to look at the total solar eclipse without eye protection?

Proper eye protection and understanding the phases of the eclipse will allow you to safely observe the event and precisely anticipate the arrival of totality - the most beautiful and dramatic phase. The duration of totality is brief, making timing crucial. The estimated times provided in this book, as well as those found in other sources, are intended for planning purposes to ensure you're in the right place at the right time. However, the exact timing of the different eclipse phases at your specific location under the path of totality can only be determined by your precise viewing position.

Here are some technical phrases you may hear regarding the phases of the eclipse:

First Contact (C1): The Moon first begins to cover the Sun.
Second Contact (C2): The Moon completely covers the Sun, marking the beginning of totality.
Third Contact (C3): The Moon begins to move away from the Sun, marking the end of totality.
Fourth Contact (C4): The Moon completely moves away from the Sun, signaling the end of the eclipse.

As a rule of thumb, during totality, viewing with your eye protection in place will show nothing, just blackness. The best way to know for sure what is happening is to use the timing information from this book or other resources as a general guide, and follow along with the actual phases of the eclipse as they occur:

Partial Phase:
As the sun is covered by the moon, the sliver of sun visible will get thinner and thinner. You will see this only through your eye protection. The partial phase begins at "first contact".

Baily's Beads:
Right before totality begins, and after totality, you might see a phenomenon known as Baily's Beads, where small beads of sunlight shine around the moon's edge. This happens because the moon's surface isn't perfectly smooth, and sunlight shines through the valleys and mountains along the moon's edge. You still need to use eye protection during this phase.

Diamond Ring:
After Baily's Beads, you might see the "diamond ring" effect, where one last spot of sunlight is visible next to a glowing ring of the sun's corona. You still need to use eye protection during this phase.

Totality:
Once the diamond ring disappears, the total eclipse, or "totality", begins. This is the only time when it's safe to look at the eclipse without eye protection. The sky will darken, and you'll see the sun's outer atmosphere, or corona, glowing around the moon. Totality can be very short, so be ready to put your glasses back on as soon as the first light of the sun appears. The "diamond ring" effect will appear again, and that ends totality. Totality begins at "second contact" and totality ends at "third contact".

End of Totality:
As soon as you see the diamond ring or Baily's Beads effect again, or any sign of the sun's bright face, you must put your eye protection back on.

Partial Phase:
The partial phase of the eclipse will begin in reverse after totality, and you must use eye protection. As the moon moves away from the sun, the visible sliver of sun will become larger and larger. You will see this only through your eye protection. This will continue until the moon no longer covers the sun, "fourth contact", and this ends the solar eclipse.

Note: It's only safe to look at the sun without eye protection during the brief moments of totality. At all other times, you must use solar viewing glasses or other appropriate eye protection to prevent serious and permanent eye damage. See the previous page for more information on where to obtain appropriate eye protection.

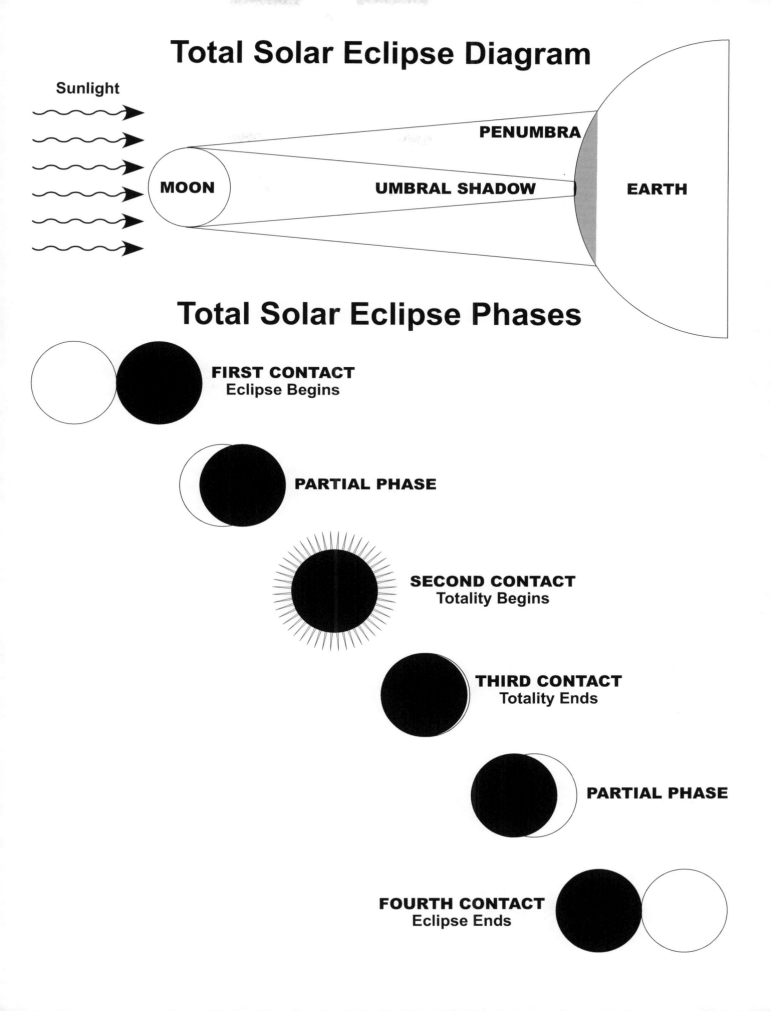

Total Solar Eclipse Diagram

Sunlight

PENUMBRA

MOON

UMBRAL SHADOW

EARTH

Total Solar Eclipse Phases

FIRST CONTACT
Eclipse Begins

PARTIAL PHASE

SECOND CONTACT
Totality Begins

THIRD CONTACT
Totality Ends

PARTIAL PHASE

FOURTH CONTACT
Eclipse Ends

Weather

How to plan around the weather and figure out where to go for clear sky on April 8, 2024
According to 18 years of data observed and compiled by NASA and presented by Jay Anderson at the *Eclipsophile.com* site, the average April cloud coverage along the central line of the eclipse ranges from 46% at Piedras Negras, Mexico / Eagle Pass, Texas, to 81% in Houlton, Maine. What I want to do here is show these figures in terms of how the daily National Weather Service forecast describes cloud cover, then show and discuss a sample of the NWS forecasts and satellite observations taken the week of April 8, 2023 to illustrate a method to track the weather in the week leading up to the eclipse, then discuss satellite image tools.

There are multiple versions of the National Weather Service cloud cover forecast characterization on *weather.gov*. Aviation forecasts use yet another, slightly different set of terms. The descriptors I have selected are from the forecast terms pages at *weather.gov*.

In table 1 below I have taken the NWS characterizations and added the decimal equivalent in the next column to match the decimal information from the NASA data. In table 2, I have used the locations and numeric designations from the NASA data, arranged them from low to high, then added the NWS sky cover expressions from the first table.

Clear/Sunny: 1/8 or less cloud coverage	00.0 - 12.5 percent coverage
Mostly Clear/Mostly Sunny: 1/8 to 3/8 cloud coverage	12.5 - 37.5 percent coverage
Partly Cloudy/Partly Sunny: 3/8 to 5/8 cloud coverage	37.5 - 62.5 percent coverage
Mostly Cloudy: 5/8 to 7/8 cloud coverage	62.5 - 87.5 percent coverage
Cloudy: 7/8 to 8/8 cloud coverage	87.5 - 100 percent coverage

Table 1: How the National Weather Service characterizes opaque cloud cover in their forecast

Average April Cloud Amount, NASA		NWS Sky Cover Expressions
Eagle Pass, TX	.46	Partly Cloudy/Partly Sunny
Kerrville, TX	.47	Partly Cloudy/Partly Sunny
Uvalde, TX	.50	Partly Cloudy/Partly Sunny
Arkadelphia, AR	.53	Partly Cloudy/Partly Sunny
Russellville, AR	.53	Partly Cloudy/Partly Sunny
Little Rock, AR	.54	Partly Cloudy/Partly Sunny
Killeen, TX	.54	Partly Cloudy/Partly Sunny
Dallas, TX	.55	Partly Cloudy/Partly Sunny
Conway, AR	.56	Partly Cloudy/Partly Sunny
Poplar Bluff, MO	.57	Partly Cloudy/Partly Sunny
Cleveland, OH	.58	Partly Cloudy/Partly Sunny
Rochester, NY	.61	Partly Cloudy/Partly Sunny
Carbondale, IL	.62	Partly Cloudy/Partly Sunny
Buffalo, NY	.65	Mostly Cloudy
Watertown, NY	.67	Mostly Cloudy
Dayton, OH	.67	Mostly Cloudy
Indianapolis, IN	.68	Mostly Cloudy
Plattsburgh, NY	.68	Mostly Cloudy
Lima, OH	.70	Mostly Cloudy
Burlington, VT	.73	Mostly Cloudy
Houlton, ME	.81	Mostly Cloudy

Table 2: Data courtesy of NASA and Jay Anderson *eclipsophile.com/2024tse/*

Applying the NWS descriptors roughly divides the eclipse track into two areas: Texas through Illinois and Indiana through Maine. Two exceptions are Cleveland, Ohio and Rochester, New York. These are called out by Jay Anderson in his Eclipsophile report at *eclipsophile.com/2024tse/* as being due to the clearing effect of Lakes Erie and Ontario that occurs slightly inland and along the south shore of these lakes.

Overall, the data points above are only averages and they don't tell the whole story. Weather is constantly changing. While the data leans toward there being fewer clouds in the southwest, and it shows that Texas is a good starting point, it's not a slam dunk for five minutes of clear sky in one place, one day in April.

So, we still need to figure out where the best prospects for clear sky are going to be in the United States on the afternoon of April 8, 2024. The method I recommend is to monitor the NWS forecast as it applies to clouds, starting 5-7 days in advance, for the area where you are planning to be. Also consider how far you'll be willing to travel in the last 12-36 hours prior to the eclipse.

On the following page is a spreadsheet, Table 3, designed to track the cloud aspects of the National Weather Service forecast for points on the eclipse track for five days leading up to the eclipse. I created this spreadsheet and tracked the results for April 3-8, 2023, as an exercise. The results are specific to this year and conditions will be completely different next year. The goal is to have a framework to determine where the best chances for a clear sky will be for your selected starting point and alternative destinations.

Most locations I selected are less than a half day drive apart, starting in the southwest and working northeastward. On April 3rd, I went to the National Weather Service forecast page at *www.weather.gov* and typed in the city and state in the upper left hand corner, and selected the city from the drop-down, to load the forecast page. From the forecast page I recorded the April 8th forecast text pertaining to clouds. I discarded information about wind, temperatures and so on. Then I moved on to the other locations in the list, recording the April 8 forecast regarding clouds. The next day, April 4th, I did the same thing and so on. Finally, observations for April 8 were made using satellite images from GOES-East at *www.star.nesdis.noaa.gov/GOES/*.

For the 21 locations I selected, the NWS forecast was correct five days out with a few minor exceptions.

First, since high-thin clouds are not opaque, the NWS counts them under sunny or mostly sunny. The eclipse can be seen through the haze of high-thin clouds, albeit with loss of many details. High-thin clouds do show up on satellite, so it is possible to anticipate them and move.

Next, the weather trend for locations to the south and west of Ennis, Texas, was cloudy. The sky was cloudy at eclipse time, but cleared off later in the day. The NWS called for partly sunny, so they were correct, just not for a specific time. A move north and east toward Clarksville would have avoided the clouds.

Finally, in Carbondale, Illinois, the forecast was mostly sunny, and it was, until the clouds started moving in. Carbondale was on the northern edge of a large weather system and caught the fringes of the system moving from the west as it passed by going southeast. A move about 60 miles north, away from that system to the northern edge of the eclipse track around Centralia, Illinois, would have dodged the clouds. The same thing happened at Vincennes, Indiana, where the sky was clear, then that large system started generating cloud cover as it passed. A move 20-30 miles north or west would have avoided the clouds.

Summary points for planning around weather:

1. Have several alternative viewing destinations, both to the southwest and northeast of your primary site.
2. Monitor the weather trend southwest and northeast of your selected location, starting 5-7 days prior.
3. The National Weather Service forecast is an accurate and valuable resource for planning around the weather.
4. High-thin clouds will let you view the eclipse through the haze with loss of detail, but not a total loss.
5. Be prepared to move the morning of the eclipse. Use current satellite images to track cloud movements.

Note: In the table below, the NWS Office column indicates the unique 3 letter code assigned to each National Weather Service office. This code can be searched online as ''weather ewx'' or used in the URL for one office. Example: FWD is the designation for the Fort Worth/Dallas, TX, NWS office and www.weather.gov/fwd is the URL for that specific office.

Table 3: Monitoring the forecast for April 8

City, State	NWS Office	National Weather Service Forecast for APR 8						OBS*
		3-Apr	4-Apr	5-Apr	6-Apr	7-Apr	8-Apr	8-Apr
Eagle Pass, TX	EWX	Cloudy	Mostly cloudy	Partly sunny	Partly sunny	Partly sunny	mostly cloudy	Cloudy
Kerrville, TX	EWX	Cloudy	Cloudy	Partly sunny	Mostly cloudy	mostly cloudy	mostly cloudy	Cloudy
Killeen, TX	FWD	Mostly cloudy	Mostly cloudy	Partly sunny	Partly sunny	Mostly cloudy	Mostly cloudy	Cloudy
Ennis, TX	FWD	Mostly cloudy	Mostly cloudy	Partly sunny	Partly sunny	Partly sunny	Partly sunny	Cloudy
Clarksville, TX	SHV	Partly sunny	Partly sunny	Partly sunny	Partly sunny	Cloudy, then gradually becoming mostly sunny	Partly sunny, then gradually becoming sunny	Sunny
Hot Springs, AR	LZK	Partly sunny	Partly sunny	Partly sunny	Partly sunny	Partly sunny	Sunny	Sunny
Morrilton, AR	LZK	Partly sunny	Partly sunny	Mostly cloudy	Partly sunny	Partly sunny	Sunny	Sunny
Ash Flat, AR	LZK	Mostly sunny	Mostly sunny	Partly sunny	Partly sunny	Partly sunny	Mostly sunny	Sunny
Poplar Bluff, MO	PAH	Mostly sunny	Mostly sunny	Partly sunny	Partly sunny	Partly sunny	Mostly sunny	High Thin
Carbondale, IL	PAH	Mostly sunny	Mostly sunny	Mostly sunny	Mostly sunny	Partly sunny	Mostly sunny	Cloudy
Vincennes, IN	IND	Mostly sunny	Sunny	Mostly sunny	Mostly sunny	Sunny	Partly cloudy this morning, then clearing	Cloudy
Franklin, IN	IND	Sunny	Sunny	Mostly sunny	Mostly sunny	Sunny	Partly cloudy this morning, then clearing.	High Thin
Muncie, IN	IND	Sunny	Sunny	Sunny	Sunny	Sunny	Mostly sunny	Sunny
Dayton, OH	ILN	Sunny	Sunny	Mostly sunny	Sunny	Sunny	Mostly sunny	High Thin
Lima, OH	IWX	Sunny	Sunny	Sunny	Sunny	Sunny	Sunny	Sunny
Cleveland, OH	CLE	Mostly sunny	Mostly sunny	Mostly sunny	Mostly sunny	sunny	mostly sunny	High Thin
Erie, PA	CLE	Mostly sunny	Mostly sunny	Mostly sunny	Sunny	Sunny	increasing clouds	High Thin
Buffalo, NY	BUF	Mostly sunny	Mostly sunny	Sunny	Mostly sunny	Sunny	Increasing clouds	High Thin
Watertown, NY	BUF	Sunny	Sunny	Mostly sunny	Sunny	Sunny	Increasing clouds	Sunny
Burlington, VT	BVT	Sunny	Mostly sunny	Mostly sunny	Sunny	Sunny	Increasing clouds	Sunny
Houlton, ME	CAR	Mostly sunny	Mostly sunny	Mostly sunny	Mostly sunny	Partly sunny	Increasing clouds	Sunny

*8-Apr Observation using GOES-East sattelite images 18:20 - 19:30 UT

Satellite Image Tools

The GOES Image viewer at *www.star.nesdis.noaa.gov/GOES/* and *Zoom.Earth*, along with the College of DuPage *weather.cod.edu/satrad* are the satellite image tools I have used to monitor the clouds over the eclipse track.

The GOES Imagery Viewer is a website hosted by NOAA / NESDIS / STAR that allows users to view images and animations of various regions captured by GOES (Geostationary Operational Environmental Satellite) satellites. The website provides imagery from both GOES-West and GOES-East satellites, covering a wide range of areas. What we want is the GOES-East - Continental U.S. (CONUS). See the image below:

Navigate to *www.star.nesdis.noaa.gov/GOES/* and select CONUS under GOES-East

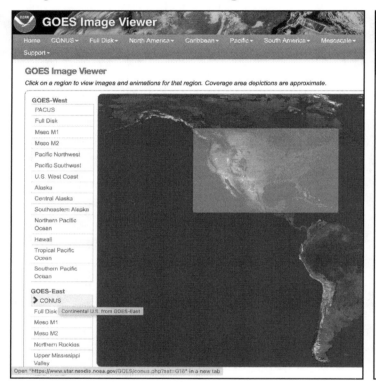 

In the GeoColor window there are various links. Animation loops opens a page where you can view an animation of the current images. Download images lets you download and view any image captured in the last ten days. Images are captured every five minutes and stored in various sizes. The other links are for different size images for the current date and time. The last link is for product documentation and goes into detail on how the images are created and how to interpret the colors shown for both day and night.

Zoom.Earth uses the same images from NOAA but has a different interface. It allows you to select the date and time on-screen, accessing images up to ten days ago. It also lets you zoom in on an area and it will add place names when zoomed in, so it provides more location details than the GOES site. Zoom Earth also lets you animate the images and has a convenient method of rocking the image forward and backward to track cloud movement.

The College of DuPage website *weather.cod.edu/satrad* has an interface to the NOAA satellite images with a convenient animation feature and access to all the different color bands and True-Color images. This website is not available on mobile.

Please see the Resources section of this book for more web links to weather and satellite resources.

Navigating the Path: Highways and Edge Cities

Texas and Oklahoma

Mobility

Texas has an extensive network of highways help you find clear sky on eclipse day. I-30 runs from Dallas, east to Sulphur Springs, Texarkana, Arkadelphia, and up to Little Rock, following the southern limit of the total eclipse. I-35 paves the way south from Dallas to Austin and San Antonio. I-10 runs northwest across the path of totality from San Antonio to Boerne, Kerrville and Junction. US-90 out of San Antonio will take you to Uvalde, Brackettville, and to the Texas/Mexico border at Del Rio. Take US-57 out of San Antonio to get to Crystal City, Carrizo Springs, and Eagle Pass. US-83 runs north/south across the path of totality from Junction to Carrizo Springs. US-281 out of San Antonio connects with Lampasas to the north where it meets US-190, joining Brady, Killeen, and Temple.

Cities near the northern and southern limits

The cities of San Antonio, Austin, and the Dallas–Fort Worth–Arlington metroplex all lie close to the edges of the total eclipse. This book lists places to eat, parks, and attractions that are well within the path of totality in these edge cities.

For San Antonio, try to be northwest of the I-410 loop, closer to State Highway 1604. Downtown San Antonio is not in the path of totality at all, so get as far north and west of there as possible.

For Austin, only the southeast corner is not in the path of totality. McKinney Falls State Park is also excluded from the show. Try to make your plans around the northern and western areas of town.

For Dallas–Fort Worth–Arlington, the northern limit of the eclipse falls just south of Denton. The farther one moves south and east toward Cleburne, Waxahachie, and Seagoville, the longer the totality phase will last. Downtown Dallas has nearly 4 minutes. On the north side, at the north end of Lewisville Lake, totality lasts only 14 seconds, at the south end of that lake it's about 2 and a half minutes.

Arkansas, Missouri and Kentucky

Mobility

In Arkansas, Interstate I-30 follows the southern limit of the total eclipse, connecting Little Rock to Arkadelphia, Texarkana, Sulphur Springs in Texas, and on to Dallas. Interstate I-40 from Little Rock will take you across the eclipse centerline and to Conway, Morrilton, Russellville, and Clarksville. US-167 continues northeastward from Little Rock, connecting with Arkansas Highway 367 and Walnut Ridge. Interstate I-40 east of Little Rock connects with I-55, creating a convenient path north to Cape Girardeau in southern Missouri.

In southern Missouri, US-60 crosses I-55 and runs east and west, connecting Poplar Bluff, Dexter, and Sikeston in Missouri to Paducah in Kentucky.

Cities near the northern and southern limits

The cities of Fort Smith, Little Rock, and Arkadelphia lie close to the edge of the total eclipse path.

For Fort Smith, to see the total eclipse at all you'll need to be east of State Highway 59 and southeast of Interstate I-49. Think about taking US-71 south toward Waldron for the longest eclipse duration.

For Arkadelphia and Little Rock, both cities are within the path of totality but are near the southern edge. If weather permits, for a longer view of totality, consider moving towards Glenwood, Hot Springs, Conway, or Morrilton, all of which are closer to the center of the total eclipse.

Illinois

Mobility

I-70 out of St. Louis will take you into the total eclipse just east of Effingham. I-57 runs from Effingham to the southern tip of Illinois. I-64 from St. Louis crosses I-57 at Mt. Vernon and continues east to just north of Evansville, Indiana. US-50 due east connects St. Louis to Salem and Olney, Illinois, and Vincennes, Indiana.

Cities near the northern and southern limits

East St. Louis, Illinois is outside the limits of the total eclipse. From St. Louis head southeast on Interstate I-64 toward Mt. Vernon or go east on US-50 to Salem or Olney. Carbondale and West Frankfort are close to the center of the eclipse and easily accessible by continuing from I-64 to I-57 south. State Highway IL-13 crosses I-57 south of West Frankfort and connects Carbondale and Marion. Effingham is right on the northern limit of the total eclipse. Head east on I-70 toward Terre Haute, Indiana, or south on I-57 toward Salem, Illinois, from Effingham for a longer duration of totality.

Indiana

Mobility

Indiana has a nice network of interstate highways facilitating travel within the path of totality. Interstates I-74 , I-65, I-69 all provide a path southward into the path of totality. I-70 cuts across the middle of the eclipse from Terre Haute, to Indianapolis, crosses the eclipse centerline near Knightstown, and continues across toward Dayton, OH. I-69 runs from Evansville in the south to Fort Wayne in the north. I-74 crosses the eclipse from Crawfordsville, southeast to Indianapolis and toward Cincinnati, Ohio. I-65 provides a north/south route from Layfayette to Indianapolis and south to Louisville, Kentucky. I-64 runs east/west across the southern tip of Indiana, connecting Evansville, Illinois to Louisville, Kentucky.

Cities near the northern and southern limits

Crawfordsville, Frankfort, Kokomo and Fort Wayne are all near the northern limit of the total eclipse. Crawfordsville, Frankfort, and Kokomo will be able to see the eclipse by moving south toward Indianapolis. For Fort Wayne, get to the south and east of I-469.

Ohio and Pennsylvania

Mobility

Ohio has a great web of interstate highways for finding a good eclipse destination. I-75 from Dayton in the south to Toledo in the north, I-71 from Columbus to Cleveland, and I-77 from Canton to Cleveland all provide north/south routes in the path of totality. US-30 runs east/west across the eclipse from Fort Wayne, Indiana to Wooster, Ohio. I-70 runs east/west above Dayton, connecting to Columbus. In the north, along Lake Erie, I-90 connects Toledo, Cleveland and Erie, Pennsylvania. In Pennsylvania, I-79 provides a direct route into the path of totality from Pittsburgh up to Erie.

Cities near the northern and southern limits

Cincinnati, Columbus, Canton, and Youngstown are all near the southern limit of the total eclipse.

For Cincinnati, I-75 north to Dayton and beyond gets you to the eclipse. For Columbus, I-71 north toward Mansfield is the direction to go. Canton, take I-77 north as far as possible. For Youngstown, Ohio State Highway 11 straight north to Ashtabula gets you as close to the centerline as you can be, or take US-422 up to Warren for a shorter duration eclipse and a shorter drive.

New York, Vermont, New Hampshire and Maine

Mobility

In New York there are several interstates to aid in moving from one area to another. I-86 in the southwest corner runs east/west above Jamestown. I-90 runs east/west along Lake Erie and connects Erie, Pennsylvania with Buffalo, Rochester, and Syracuse. I-90 also provides a route into the path of totality from Albany, Schenectady, and Utica. I-81 connects Syracuse to Watertown. I-87 connects Plattsburgh to Albany in eastern New York.

In Vermont, I-89 connects St. Albans City, Burlington and the capitol, Montpelier. I-91 in eastern Vermont is the north/south corridor connecting Newport, Lyndon and St. Johnsbury .

The eclipse crosses extreme northern New Hampshire where US-3 runs north/south along the border with Vermont, connecting Colebrook and Northumberland.

In Maine, I-95 provides a route into the total eclipse path and crosses US-1 near Houlton. US-1 runs north/south, connecting Caribou, Houlton and points south.

Cities near the northern and southern limits

In New York, Olean, Syracuse and Rome are near the southern boundary of the eclipse. Olean can head west to Jamestown or north to the Buffalo area. Syracuse is within the path of totality but just barely, with the south east part of I-481 being right on the edge of the eclipse path. The farther north you go in Syracuse towards Fulton or Oswego, the longer the eclipse will last for you, providing that the weather is clear. For Rome, move north towards Watertown to see the eclipse. This book lists places to eat, parks, and attractions that are all well within the path of totality in these edge cities.

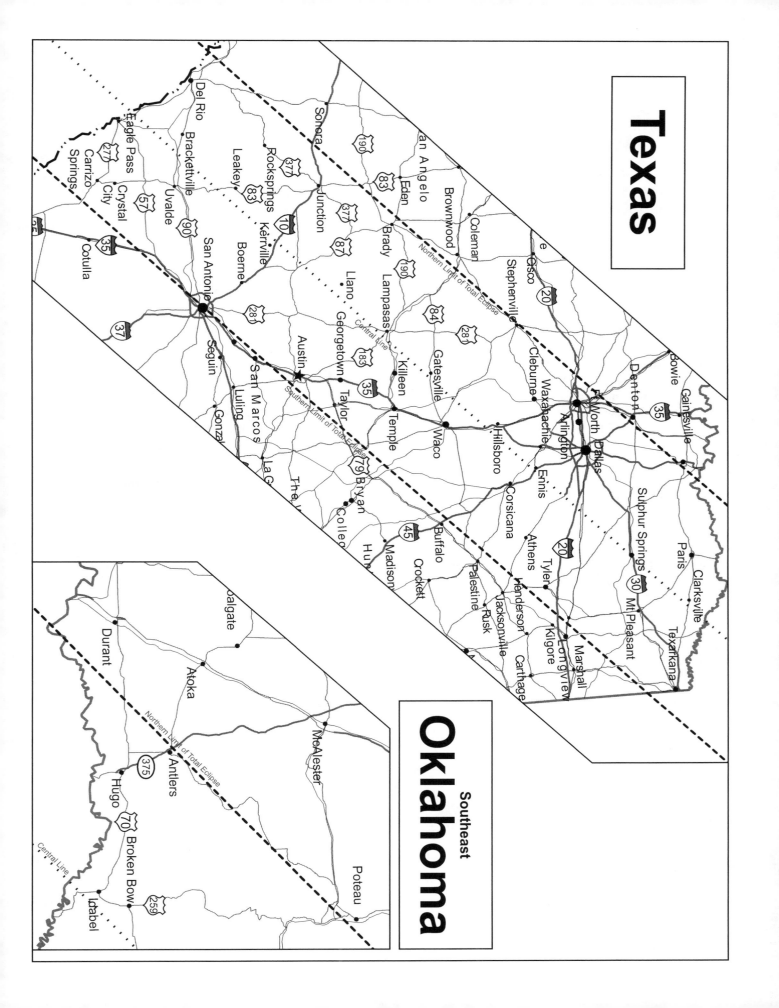

Featured Places to Eat

Billy Gene's Restaurant in Kerrville offers a menu that captures the flavors of the Texas Hill Country. Appetizers include options like chips and salsa, fried pickles, and onion rings. The restaurant is known for its half-pound burgers, such as the Billy Gene Burger, and Swiss Mushroom Burger, as well as sandwiches such as pulled pork. Kerrville is very close to the center line for this eclipse and visitors can experience well over 4 minutes of totality. Located at 1489 Junction Hwy, (830) 895-7377

Brasão Brazilian Steakhouse in San Antonio specializes in Brazilian cuisine. They have a salad bar with exotic cheeses, but the main focus of the menu is the meats, which are grilled over an open fire. Enjoy Picanha and Fraldinha, along with Ribeye, Alcatra with cheese, Filet Mignon, succulent Beef Ribs and many other delights. You'll also be treated to over two minutes of totality in San Antonio. Located at 19210 I-10, (210) 233-6868

Galaxy B&G in Killeen is a space-themed restaurant that offers an enticing and playful menu for breakfast, lunch, and dinner. For breakfast, diners can enjoy pancakes, French toast, omelets, and breakfast tacos. For lunch and dinner, burgers win with a variety of toppings. Try "The Darkside," a burger topped with bacon, cheese, lettuce, tomato, onions, pickles, and a fried egg. Prepare your snacks and witness over 4 minutes of cosmic glory in Killeen during the total solar eclipse, beginning at 1:36. Located at 104 W Veterans Memorial Blvd, (254) 213-9888

Habesha Ethiopian Restaurant and Bar in Austin has a variety of Ethiopian dishes and drinks. Their menu includes appetizers such as Sambusa (stuffed pastry) and Qategna (flatbread), entrees such as Kitfo (spiced beef) and Doro Wat (chicken stew), and vegetarian options. You'll be able to catch over 2 minutes of totality from this area. Try out something new and enjoy the eclipse! Located at 6019 N Interstate Hwy 35, (512) 358-6839

Limericks Sandwich Junction in Ennis offers an appetizing selection of sandwiches. The deli-style Weinberger Reuben, packed with hot corned beef, pastrami, Swiss cheese, sauerkraut, and Russian dressing on grilled marble rye, is sure to delight. They have hot pastrami, subs, Cubano sandwiches, cheesesteak and many other satisfying options. Ennis is right on the center line for this eclipse and you can see well over 4 minutes of totality. Located at 213 W Ennis Ave Ste 100, (903) 405-1995

Terry Black's BBQ in Dallas is a premiere destination for legendary Texas barbecue featuring sliced brisket, pork ribs, chopped beef, original sausage, and beef ribs. They offer an array of savory sides such as mac-n-cheese, pinto beans, Mexican rice, and baked potato salad, as well as deserts. The sky's greatest show lasts for over 3 minutes in Dallas; get your BBQ and enjoy the show! Located at 3025 Main St, (469) 399-0081

Eclipse Track Notes - TX - OK

Cities near the northern and southern limits

The cities of San Antonio, Austin, and the Dallas–Fort Worth–Arlington metroplex all lie close to the edges of the total eclipse. If viewing around San Antonio or Austin, try to make your plans around the northern and western areas of town. This book lists restaurant locations that are all well within the path of totality in these edge cities. For Dallas–Fort Worth–Arlington, the northern limit of the eclipse falls just south of Denton. The farther one moves towards Cleburne, Waxahachie, and Seagoville, the longer the totality phase will last.

Mobility

The interstate highways in Texas provide some mobility options. I-30 runs from Dallas east to Sulphur Springs, Texarkana, Arkadelphia and up to Little Rock, following the southern limit of the total eclipse. I-35 will take you south from Dallas to Austin and San Antonio. I-10 runs northwest from San Antonio to Boerne, Kerrville and Junction. US-90 out of San Antonio will take you to Uvalde, Brackettville, and to the Texas/Mexico border at Del Rio.

Eagle Pass, TX
Totality Starts at: 1:27pm CDT
Totality Lasts for: 4:22
Totality Ends at: 1:27pm CDT
NWS Office: www.weather.gov/ewx
www.eaglepasstx.us

El Zarape Grill
2328 El Indio Hwy, (830) 757-5364
Mexican, Eagle Pass
Eclipse Day Monday, 7am - 2pm

Felipe's Hot Dogs
1591 El Indio Hwy, (830) 757-6656
Mexican, Eagle Pass
Eclipse Day Eclipse Day Monday, 9am - 8pm

La Casita Restaurant
403 Memo Robinson Dr, (830) 757-4084
Mexican, Eagle Pass
Eclipse Day Monday, 6am - 2pm

Miguel y Miguel
1592 El Indio Hwy, (830) 758-0383
Mexican, Eagle Pass
Eclipse Day Monday, 7am - 3pm

Mi Ranchito Mexican Restaurant
1720 Del Rio Blvd, (830) 325-3663
Mexican, Eagle Pass
Eclipse Day Monday, 7am - 2pm

New China Buffet
410 S Texas # E, (830) 757-5777
Chinese, Eagle Pass
Eclipse Day Monday, 11am - 8:30pm

Parrilla De San Miguel
408 S Texas Dr, (830) 757-3100
Mexican, Eagle Pass
Eclipse Day Monday, 11am - 10pm

Rodee's Country Fried Chicken
1910 El Indio Hwy, (830) 757-8888
American, Eagle Pass
Eclipse Day Monday, 11am - 7:30pm

Skillets Restaurant
2505 E Main St, (830) 773-7263
American, Eagle Pass
Eclipse Day Monday, 7am - 10pm

Yopo's
455 S Bibb Ave, (830) 773-0983
American, Eagle Pass
Eclipse Day Monday, 9:30am - 10pm

Del Rio, TX
Totality Starts at: 1:28pm CDT
Totality Lasts for: 3:24
Totality Ends at: 1:31pm CDT
NWS Office: www.weather.gov/ewx
www.cityofdelrio.com

830 Kitchen Restaurant
301 Ave B, (830) 308-3806
American, Del Rio
Eclipse Day Monday, 11am - 7pm

Benny's Cafe
650 S Main St, (830) 774-3176
American, Del Rio
Eclipse Day Monday, 7:30am - 3pm

Border One Stop
2301 State Loop 239, (830) 778-2260
Mexican, Del Rio
Eclipse Day Monday, 7am - 8pm

Dona Elvira Mexican Restaurant
206 Mary Lou Dr, (830) 774-2049
Mexican, Del Rio
Eclipse Day Monday, 7am - 2pm

El Patio Del Rio
1206 E Gibbs St, (830) 774-8080
Mexican, Del Rio
Eclipse Day Monday, 6am - 8pm

Los Beto's Restaurant
1912 Veterans Blvd A, (830) 734-2960
Mexican, Del Rio
Eclipse Day Monday, 7am - 5pm

Lou's Woodfire Pizza
2409 Veterans Blvd STE 3, (830) 422-2150
Pizza, Del Rio
Eclipse Day Monday, 11am - 10pm

Mexico Tipico Mexican Food
1302 Las Vacas St, (830) 422-2132
Mexican, Del Rio
Eclipse Day Monday, 7am - 3pm

Never gaze at the sun without eye protection. Only remove eye protection during 100% totality.

Mi Tierra Linda Restaurant
909 Dr Fermin Calderon Blvd, (830) 422-2224
Mexican, Del Rio
Eclipse Day Monday, 7am - 2pm

Molcajetes Restaurant
808 Ave G, (830) 309-7060
Mexican, Del Rio
Eclipse Day Monday, 10am - 8pm

Patty's Restaurant
1750 Veterans Blvd, (830) 320-8019
American, Del Rio
Eclipse Day Monday, 7am - 3pm

Rudy's "Country Store" and Bar-B-Q
330 Braddie Dr, (830) 774-0784
BBQ, Del Rio
Eclipse Day Monday, 7am - 10pm

Salas Better Burger
913 E Ogden St, (830) 775-0051
American, Del Rio
Eclipse Day Monday, 10am - 9pm

Skillet's Restaurant
2003 Veterans Blvd, (830) 775-6060
American, Del Rio
Eclipse Day Monday, 6am - 10pm

Brackettville, TX
Totality Starts at: 1:28pm CDT
Totality Lasts for: 4:15
Totality Ends at: 1:33pm CDT
NWS Office: www.weather.gov/ewx
www.thecityofbrackettville.com

Julie's Place
US-90, (830) 563-9511
American, Brackettville
Eclipse Day Monday, 10am - 2:30 pm, 5–8:30pm

Manita's Restaurant
301 Spring St, (830) 563-9557
Mexican, Brackettville
Eclipse Day Monday, 10am - 2 pm, 5–9pm

The Pizza Outpost
302 E Military Hwy, (830) 867-4008
Pizza, Brackettville
Eclipse Day Monday, 12–9pm

Ziggy's Roadside BBQ
120 U.S. Hwy 90 Street, (830) 563-9293
BBQ, Brackettville
Eclipse Day Monday, 6am - 5pm

Carrizo Springs, TX
Totality Starts at: 1:29pm CDT
Totality Lasts for: 2:51
Totality Ends at: 1:31pm CDT
NWS Office: www.weather.gov/ewx
www.cityofcarrizo.org

5 D GRILL & LOUNGE
2525 US-83, (830) 219-0118
American, Carrizo Springs
Eclipse Day Monday, 11am - 10pm

Amigo's Steakhouse
2207 US-83, (830) 876-3778
Steak, Carrizo Springs
Eclipse Day Monday, 11am - 10pm

MJ's Mexican Restaurant
317 E Nopal St, (830) 322-5141
Mexican, Carrizo Springs
Eclipse Day Monday, 6am - 2pm

Miguelito's Mexican Grill and Cantina
101 Petry Pl, (830) 876-3565
Mexican, Carrizo Springs
Eclipse Day Monday, 11am - 9pm

Rosita's Restaurant
604 N 1st St, (830) 876-3825
American, Carrizo Springs
Eclipse Day Monday, 6am - 9pm

Stars Drive-in
1906 N 1st St, (830) 876-9568
American, Carrizo Springs
Eclipse Day Monday, 10am - 11pm

Tacos El Rey
701 N 1st St, (830) 325-8687
Mexican, Carrizo Springs
Eclipse Day Monday, 12–10pm

Tina's Tacos
2824 US-83, (830) 255-6575
Mexican, Carrizo Springs
Eclipse Day Monday, 5am - 2pm

Crystal City, TX
Totality Starts at: 1:29pm CDT
Totality Lasts for: 3:16
Totality Ends at: 1:32pm CDT
NWS Office: www.weather.gov/ewx
www.crystalcitytx.org

Anthony's Restaurant
2110 US-83, (830) 854-2032
Mexican, Crystal City
Eclipse Day Monday, 11am - 9pm

Arturo's Tacos
1524 N 1st Ave, (830) 854-2200
Mexican, Crystal City
Eclipse Day Monday, 11:30am - 10pm

Miguelito's Mexican Grill
1023 N 1st Ave, (830) 374-3461
Mexican, Crystal City
Eclipse Day Monday, 10am - 9pm

Tasty Taco Drive Inn
912 E Crockett St, (830) 374-2560
Mexican, Crystal City
Eclipse Day Monday, 7am - 2pm

Yolie's Steakhouse & Mexican
US-83, (830) 374-2772
Mexican, Crystal City
Eclipse Day Monday, 8am - 8pm

Uvalde, TX
Totality Starts at: 1:29pm CDT
Totality Lasts for: 4:16
Totality Ends at: 1:33pm CDT
NWS Office: www.weather.gov/ewx
www.uvaldetx.gov

Billy Bob's Hamburgers
2342 E Main St, (830) 261-5694
American, Uvalde
Eclipse Day Monday, 10am - 10pm

Broadway 830
100 E Main St, (830) 900-7076
Pizza, Uvalde
Eclipse Day Monday, 11am - 10pm

El Herradero de Jalisco
224 W Main St, (830) 278-3600
Mexican, Uvalde
Eclipse Day Monday, 6:30am - 10pm

Hangar 6 Air Cafe
249 Airport, Uvalde Rd, (830) 900-3113
American, Uvalde
Eclipse Day Monday, 11am - 8pm

Jack's Steak House
2500 E Main St, (830) 278-9955
Steak, Uvalde
Eclipse Day Monday, 11am - 9pm

Julio's Grill
501 S Getty St, (830) 591-2099
Mexican, Uvalde
Eclipse Day Monday, 6am - 2pm

La Charreada Mexican Restaurant
2042 E Main St, (830) 278-8998
Mexican, Uvalde
Eclipse Day Monday, 8am - 3pm

Oasis Outback BBQ & Grill
2900 E Main St, (830) 278-4000
BBQ, Uvalde
Eclipse Day Monday, 10:30am - 3pm

PHO LONG
2000 E Main St Suite B, (830) 261-5754
Vietnamese, Uvalde
Eclipse Day Monday, 9am - 8pm

Sunrise Restaurant
510 W Main St, (830) 278-6100
American, Uvalde
Eclipse Day Monday, 6am - 2pm

Taco Bliss Express
735 W Main St, (830) 900-6174
Mexican, Uvalde
Eclipse Day Monday, 6am - 7:30pm

The Local Fix
2001 E Main St, (830) 900-7183
Breakfast, Uvalde
Eclipse Day Monday, 7am - 3pm

Town House Restaurant
2105 E Main St, (830) 278-2428
American, Uvalde
Eclipse Day Monday, 11am - 9pm

Never gaze at the sun without eye protection. Only remove eye protection during 100% totality.

Leakey, TX
Totality Starts at: 1:30pm CDT
Totality Lasts for: 4:25
Totality Ends at: 1:34pm CDT
NWS Office: www.weather.gov/ewx
www.co.real.tx.us

Gypsy Sally's
373 US-83, (830) 260-1001
American, Leakey
Eclipse Day Monday, 11am - 9pm

Lala's Mexican Restaurant
491 US-83, (830) 232-4572
Mexican, Leakey
Eclipse Day Monday, 6:30am - 2pm

Leakey BBQ
US-83, (832) 622-6366
BBQ, Leakey
Eclipse Day Monday, 11am - 8pm

Mama Chole's
234 US-83, (830) 232-6111
Mexican, Leakey
Eclipse Day Monday, 11am - 9pm

Mill Creek Cafe
849 US-83 South, (830) 232-4805
American, Leakey
Eclipse Day Monday, 7am - 3pm

Vinny's Italian Restaurant and Pizzeria
311 US-83, (830) 232-4420
Pizza, Leakey
Eclipse Day Monday, 12–2 pm, 4–8pm

Rocksprings, TX
Totality Starts at: 1:30pm CDT
Totality Lasts for: 3:22
Totality Ends at: 1:34pm CDT
NWS Office: www.weather.gov/ewx
cityofrockspringstx.com

Jail House Grill & Bar
108 W Austin St, (830) 683-3366
American, Rocksprings
Eclipse Day Monday, 11am - 2 pm, 5–9pm

King Burger Drive Inn
102 N State St, (830) 683-4127
American, Rocksprings
Eclipse Day Monday, 7:45am - 8:45pm

Lotus Thai Cafe
103 W Main St, (830) 683-3711
Thai, Rocksprings
Eclipse Day Monday, 11am - 2pm

Kerrville, TX
Totality Starts at: 1:32pm CDT
Totality Lasts for: 4:23
Totality Ends at: 1:36pm CDT
NWS Office: www.weather.gov/ewx
www.kerrvilletx.gov

Bella Sera of Kerrville, Inc.
2124 Sidney Baker St, (830) 257-2661
Italian, Kerrville
Eclipse Day Monday, 11am - 9pm

Billy Gene's Restaurant
1489 Junction Hwy, (830) 895-7377
American, Kerrville
Eclipse Day Monday, 11am - 8:30pm

Black Jack BBQ
1700 Water St, (830) 955-5572
BBQ, Kerrville
Eclipse Day Monday, 7am - 7pm

Branding Iron restaurant
2033 Sidney Baker St, (830) 257-4440
American, Kerrville
Eclipse Day Monday, 6:30am - 2pm

Buddy's Good Food
3324 Junction Hwy, (830) 353-1411
American, Kerrville
Eclipse Day Monday, 11am - 3pm

Del Norte Restaurant
710 Junction Hwy, (830) 257-3337
American, Kerrville
Eclipse Day Monday, 7am - 2pm

Dickeys Barbecue Pit
881 Junction Hwy, (830) 792-0757
BBQ, Kerrville
Eclipse Day Monday, 11am - 8:50pm

El Brasero de Jalisco
843 Junction Hwy, (830) 896-0926
Mexican, Kerrville
Eclipse Day Monday, 6am - 9pm

El Sombrero De Jalisco
303 Sidney Baker St S, (830) 257-8263
Mexican, Kerrville
Eclipse Day Monday, 7am - 10pm

Francisco's Restaurant
201 Earl Garrett St, (830) 257-2995
Mexican, Kerrville
Eclipse Day Monday, 11am - 2pm

Los Jimadores Mexican Grill & Bar
1550 Junction Hwy, (830) 315-2551
Mexican, Kerrville
Eclipse Day Monday, 11am - 9pm

Mamacita's Restaurant
215 Junction Hwy, (830) 895-2441
Mexican, Kerrville
Eclipse Day Monday, 11am - 9:30pm

Mary's Tacos
1616 Broadway, (830) 895-7474
Mexican, Kerrville
Eclipse Day Monday, 6am - 1:30pm

Rails A Cafe At the Depot
615 E Schreiner St, (830) 257-3877
American, Kerrville
Eclipse Day Monday, 11am - 8pm

Ringo's On The River
1421 Junction Hwy, (830) 285-8638
American, Kerrville
Eclipse Day Monday, 11am - 11pm

Save Inn Restaurant
1806 Sidney Baker St, (830) 257-7484
American, Kerrville
Eclipse Day Monday, 7am - 2pm

Taqueria Jalisco
2190 Junction Hwy, (830) 257-0606
Mexican, Kerrville
Eclipse Day Monday, 6am - 9pm

Junction, TX
Totality Starts at: 1:32pm CDT
Totality Lasts for: 3:05
Totality Ends at: 1:35pm CDT
NWS Office: www.weather.gov/sjt
www.junctiontexas.com

Cooper's Bar-B-Q & Grill
2423 N Main St, (325) 446-8664
BBQ, Junction
Eclipse Day Monday, 10am - 8pm

Cowboy Grill
2341 N Main St, (325) 446-2775
American, Junction
Eclipse Day Monday, 11:30am - 9:30pm

Gloria's Gonzales Cafe
1106 Main St, (325) 446-4202
American, Junction
Eclipse Day Monday, 11am - 9pm

Isaack Restaurant
1606 Main St, (325) 446-2629
American, Junction
Eclipse Day Monday, 6am - 9pm

Junction Burger Company
1907 Main St, (325) 446-2695
American, Junction
Eclipse Day Monday, 11am - 9pm

Lum's BBQ
2031 Main St, (325) 446-3541
BBQ, Junction
Eclipse Day Monday, 10:30am - 9pm

Brady, TX
Totality Starts at: 1:34pm CDT
Totality Lasts for: 2:02
Totality Ends at: 1:36pm CDT
NWS Office: www.weather.gov/sjt
www.bradytx.us

Brady's Restaurant
503 S Bridge St, (325) 597-9990
American, Brady
Eclipse Day Monday, 11am - 9pm

Never gaze at the sun without eye protection. Only remove eye protection during 100% totality.

Cattleman's BBQ
2010 S Bridge St, (325) 597-2999
BBQ, Brady
Eclipse Day Monday, 11am - 8pm

Medina's Restaurant
701 S Bridge St, (325) 597-1847
Mexican, Brady
Eclipse Day Monday, 5–11 AM

México City Cafe
706 N Bridge St, (325) 597-0442
Mexican, Brady
Eclipse Day Monday, 8am - 8pm

Mi Familia
100 S Church St, Brady, (325) 597-1037
Mexican, Brady
Eclipse Day Monday, 11am - 2 pm, 5–9pm

Mi Pueblo 2 Mexican Restaurant
2035 S Bridge St, (325) 597-2693
Mexican, Brady
Eclipse Day Monday, 7:30am - 9pm

Mr. China
300 S Bridge St, (325) 597-2141
Chinese, Brady
Eclipse Day Monday, 11am - 9:30pm

Sandy's Kitchen and Catering
2105 S Bridge St, (325) 240-9490
American, Brady
Eclipse Day Monday, 6am - 2pm

San Antonio, TX
Totality Starts at: 1:33pm CDT
Totality Lasts for: ~ 2:31 - Longer NW
Totality Ends at: 1:35pm CDT
NWS Office: www.weather.gov/ewx
www.sanantonio.gov

Amiga Cafe
5309 Wurzbach Rd #115, (210) 255-8003
Mexican, San Antonio
Eclipse Day Monday, 6am - 9pm

Blanco BBQ
13259 Blanco Rd, (210) 251-2602
BBQ, San Antonio
Eclipse Day Monday, 11am - 9pm

Brasão Brazilian Steakhouse
19210 I-10, (210) 233-6868
Brazilian, San Antonio
Eclipse Day Monday, 11am - 2 pm, 5–10pm

El Buen Gusto Mexican Cafe
7709 Tezel Rd, (210) 681-1773
Mexican, San Antonio
Eclipse Day Monday, 7am - 2:30pm

El Chaparral Mexican Restaurant - Helotes
15103 Bandera Rd, (210) 695-8302.
Mexican, Helotes
Eclipse Day Monday, 11am - 9pm

El Tequila Mexican Restaurant & Bar
10910 Marbach Rd #101, (210) 451-9373
Mexican, San Antonio
Eclipse Day Monday, 6am - 11pm

Elizabeth's Mexican Restaurant
5251 Timberhill Dr #206, (210) 520-2280
Mexican, San Antonio
Eclipse Day Monday, 6am - 2pm

Henry's Puffy Tacos
6030 Bandera Rd, (210) 647-8339
Mexican, San Antonio
Eclipse Day Monday, 10:30am - 9pm

Hon Machi Korean BBQ
19186 Blanco Rd # 103, (210) 670-7128
Korean BBQ, San Antonio
Eclipse Day Monday, 11am - 10pm

Lin's International Buffet
7915 W Loop 1604 N, (210) 476-5027
Asian & Seafood, San Antonio
Eclipse Day Monday, 11am - 9:30pm

Little Italy Restaurant & Pizzeria
824 Afterglow St, (210) 349-2060
Italian, San Antonio
Eclipse Day Monday, 11am - 9pm

Los Azulejos Restaurant
2267 NW Military Hwy Suite 101, (210) 281-4500
Mexican, San Antonio
Eclipse Day Monday, 11am - 9pm

Pappadeaux Seafood Kitchen
15715 I-10, (210) 641-1171
Seafood, San Antonio
Eclipse Day Monday, 11am - 9pm

Panfila Cantina
22250 Bulverde Rd #114, (210) 455-0702
Mexican, San Antonio
Eclipse Day Monday, 10am - 9pm

Picnikins Patio Cafe
5811 University Heights Blvd, (210) 236-5134
American, San Antonio
Eclipse Day Monday, 11am - 3:30pm

Rosario's Mexican Restaurant y Lounge
9715 San Pedro Ave, (210) 481-4100
Mexican, San Antonio
Eclipse Day Monday, 11am - 4pm

Sweet Paris Creperie & Cafe
15900 La Cantera Pkwy, (210) 561-4452
Creperie, San Antonio
Eclipse Day Monday, 9am - 8pm

Thai Bistro & Sushi
5999 De Zavala Rd, (210) 558-6707
Thai, San Antonio
Eclipse Day Monday, 11am - 9:30pm

Thai Hut Restaurant
9902 Potranco Rd, (210) 647-1022
Thai, San Antonio
Eclipse Day Monday, 11am - 9pm

The Longhorn Cafe
17625 Blanco Rd, (210) 492-0301
American, San Antonio
Eclipse Day Monday, 11am - 8:30pm

The Magnolia Pancake Haus
10333 Huebner Rd, (210) 496-0828
Breakfast, San Antonio
Eclipse Day Monday, 7am - 2pm

Boerne, TX
Totality Starts at: 1:32pm CDT
Totality Lasts for: 3:34
Totality Ends at: 1:36pm CDT
NWS Office: www.weather.gov/ewx
www.ci.boerne.tx.us

Bella Sera of Boerne
812 N Main St, (830) 816-5577
Italian, Boerne
Eclipse Day Monday, 11am - 9pm

Boerne Taco House
470 S Main St Suite 200, (830) 331-2874
Mexican, Boerne
Eclipse Day Monday, 11am - 8pm

Bumdoodlers Lunch Co
929 N Main St, (830) 249-8826
American, Boerne
Eclipse Day Monday, 10am - 6pm

Casa Amaya Pupuseria & Taqueria
109 Waterview Pkwy Ste 103, (830) 331-4190
Mexican, Boerne
Eclipse Day Monday, 9am - 5pm

Compadres Hill Country Cocina LLC
209 Lohmann St, (830) 331-2198
Mexican, Boerne
Eclipse Day Monday, 11am - 2pm

Dog & Pony
1481 S Main St, (830) 816-7669
American, Boerne
Eclipse Day Monday, 11am - 8pm

Guadalajara Mexican Grill
1234 S Main St, (830) 249-1467
Mexican, Boerne
Eclipse Day Monday, 6:30am - 1:30pm

Hungry Horse Restaurant and Catering
109 Saunders St, (830) 816-8989
American, Boerne
Eclipse Day Monday, 11am - 9pm

Las Guitarras Mexican Restaurant
911 S Main St, (830) 331-8787
Mexican, Boerne
Eclipse Day Monday, 11am - 9pm

Longhorn Cafe
369 S Esser Rd, (830) 331-4011
American, Boerne
Eclipse Day Monday, 11am - 8:30pm

Mary's Tacos
518 E Blanco Rd, (830) 249-7474
Mexican, Boerne
Eclipse Day Monday, 6am - 1:30pm

Nico's Gourmet Burgers
109 Waterview Pkwy Ste 104, (830) 431-6836
American, Boerne
Eclipse Day Monday, 11:30am - 5pm

Never gaze at the sun without eye protection. Only remove eye protection during 100% totality.

Sauced Wing Bar
215 W Bandera Rd #101, (830) 331-4398
Wings, Boerne
Eclipse Day Monday, 11am - 9pm

Shang Hai Chinese Restaurant
430 W Bandera Rd # 1, (830) 249-4982
Chinese, Boerne
Eclipse Day Monday, 11am - 3 pm, 4:30–9:30pm

Taqueria Reyna Tapatia
1000 N Main St, (830) 331-1076
Mexican, Boerne
Eclipse Day Monday, 6am - 8:30pm

The Dodging Duck Brewhaus & Restaurant
402 River Rd, (830) 248-3825
American, Boerne
Eclipse Day Monday, 11am - 9pm

Llano, TX
Totality Starts at: 1:34pm CDT
Totality Lasts for: 4:23
Totality Ends at: 1:38pm CDT
NWS Office: www.weather.gov/ewx
www.cityofllano.com

Chap's 29 Cafe and Grill
910 W Young St, (325) 307-2676
American, Llano
Eclipse Day Monday, 6am - 3pm

Cooper's Old Time Pit Bar-B-Que
604 W Young St, (325) 247-5713
BBQ, Llano
Eclipse Day Monday, 11am - 8pm

La Hacienda De Jalisco
303 E Young St, (325) 247-1022
Mexican, Llano
Eclipse Day Monday, 6am - 9pm

Llano's Hungry Hunter
702 W Young St, (325) 247-4236
American, Llano
Eclipse Day Monday, 6am - 2pm

Trailblazer Grille - Llano
109 W Main St, (325) 248-0500
American, Llano
Eclipse Day Monday, 11am - 8pm

Austin, TX
Totality Starts at: 1:36pm CDT
Totality Lasts for: ~ 2:18 - Longer NW
Totality Ends at: 1:38pm CDT
NWS Office: www.weather.gov/ewx
www.austintexas.gov

888 Pan Asian Restaurant
2400 E Oltorf St #1A, (512) 448-4722
Asian, Austin
Eclipse Day Monday, 12–11pm

Aleidas Restaurant
2011 Little Elm Trail suite 106, (512) 551-2104
Latin American, Austin
Eclipse Day Monday, 9am - 8pm

Bartlett's Restaurant
2408 W Anderson Ln., (512) 451-7333
Steak, Austin
Eclipse Day Monday, 11am - 9pm

Casamigos Tex-Mex
15609 Ronald Reagan Blvd, (512) 456-7895
Tex-Mex, Leander
Eclipse Day Monday, 11am - 9pm

Ebisu Japanese Restaurant
13376 Research Blvd Ste 400, (512) 243-5554
Japanese, Austin
Eclipse Day Monday, 11am - 9:30pm

El Arroyo
1624 W 5th St, (512) 474-1222
Mexican, Austin
Eclipse Day Monday, 11am - 10pm

Gino's Italian Restaurant and Pizza
1701 S Mays St, (512) 218-9922
Italian, Austin
Eclipse Day Monday, 11am - 2 pm, 5–9pm

Habanero Cafe
501 W Oltorf St, (512) 416-0443
Mexican, Austin
Eclipse Day Monday, 7am - 3pm

Habesha Ethiopian restaurant and bar
6019 N Interstate Hwy 35, (512) 358-6839
Ethiopian, Austin
Eclipse Day Monday, 11am - 9pm

Jack Allen's Kitchen Oak Hill
7720 State Hwy 71 West, (512) 852-8558
American, Austin
Eclipse Day Monday, 11am - 9pm

Ken's Tacos
9408 Dessau Rd, (512) 837-9370
Mexican, Austin
Eclipse Day Monday, 6am - 2pm

Lazeez Mediterranean Cafe
6812 N FM 620, (512) 215-9356
Mediterranean, Austin
Eclipse Day Monday, 11am - 9pm

Lebowski's Grill at Highland Lanes
8909 Burnet Rd, (512) 419-7166
American, Austin
Eclipse Day Monday, 12–10pm

Phoebe's Diner - Oltorf
533 W Oltorf St, (512) 883-9682
Breakfast, Austin
Eclipse Day Monday, 7am - 3pm

Poke House Austin
11150 Research Blvd #216, (512) 291-6986
Hawaiian, Austin
Eclipse Day Monday, 11am - 9pm

Santorini Cafe
11800 N Lamar Blvd, (512) 833-6000
Greek, Austin
Eclipse Day Monday, 11am - 10pm

Secret Taco Stand
7612 Bluff Springs Rd
Mexican, Austin
Eclipse Day Monday, 7am - 2 pm, 4 pm–2 AM

Smokey Mo's BBQ
6001 W Parmer Ln, (512) 918-0002
BBQ, Austin
Eclipse Day Monday, 7am - 9pm

Tx shawarma
601 W Live Oak St, (512) 401-3638
Mediterranean, Austin
Eclipse Day Monday, 11am - 9:30pm

Taco Flats-Burnet Road
5520 Burnet Rd #101, (512) 284-8352
Mexican, Austin
Eclipse Day Monday, 10am - 10pm

Taylor, TX
Totality Starts at: 1:37pm CDT
Totality Lasts for: 1:51
Totality Ends at: 1:38pm CDT
NWS Office: www.weather.gov/ewx
www.ci.taylor.tx.us

El Corral Lozano
300 W 2nd St, (512) 352-3728
Mexican, Taylor
Eclipse Day Monday, 7am - 3pm

Fajitas On Wheels
401 W 3rd St, (512) 352-2326
Mexican, Taylor
Eclipse Day Monday, 8am - 1pm

Gonzalez Taco's
800 E M.L.K. Jr Blvd, (512) 365-6781
Mexican, Taylor
Eclipse Day Monday, 7am - 1:30pm

Lucky Duck Cafe
220 E 4th St, (512) 352-8777
American, Taylor
Eclipse Day Monday, 11am - 8pm

Mariachis de Jalisco
3100 N Main St, (512) 352-1700
Mexican, Taylor
Eclipse Day Monday, 6am - 10pm

Masfajitas
1905 N Main St, (512) 352-9292
Mexican, Taylor
Eclipse Day Monday, 11am - 9pm

Mr Gatti's Pizza
2708 N Main St, (512) 365-2222
Pizza, Taylor
Eclipse Day Monday, 11am - 9pm

New York Pizza Pasta
1701 W 2nd St, (512) 352-6000
Italian, Taylor
Eclipse Day Monday, 11am - 9pm

Plowman's Kitchen
305 W 9th St, (512) 352-0055
American, Taylor
Eclipse Day Monday, 9am - 8pm

Never gaze at the sun without eye protection. Only remove eye protection during 100% totality.

Rojas Tacos
400 S Main St, (512) 945-2750
Mexican, Taylor
Eclipse Day Monday, 6:30am - 1pm

Zapata's Mexican Restaurant
1808 W 2nd St, (512) 352-3995
Mexican, Taylor
Eclipse Day Monday, 5:30am - 2pm

Georgetown, TX
Totality Starts at: 1:36pm CDT
Totality Lasts for: 3:16
Totality Ends at: 1:39pm CDT
NWS Office: www.weather.gov/ewx
georgetown.org

600 Degrees Pizzeria
124 E 8th St, (512) 943-9272
Pizza, Georgetown
Eclipse Day Monday, 11am - 9pm

Asian Cuisine Restaurant
1203 Leander Rd, (512) 763-1206
Thai, Chinese & Vietnamese, Georgetown
Eclipse Day Monday, 11am - 9pm

Blue Corn Harvest Bar & Grill
212 W 7th St #105, (512) 819-6018
Southwest, Georgetown
Eclipse Day Monday, 11am - 9pm

Dos Salsas
1104 S Main St, (512) 930-2343
Tex-Mex, Georgetown
Eclipse Day Monday, 8am - 10pm

El Charrito
302 S Austin Ave, (512) 930-9137
Mexican, Georgetown
Eclipse Day Monday, 7am - 5pm

El Tio Nacho Taqueria
112 Woodmont Dr, (512) 831-1581
Mexican, Georgetown
Eclipse Day Monday, 6:30am - 8:30pm

Hat Creek Burger Company
201 San Gabriel Village Blvd, (512) 943-8258
American, Georgetown
Eclipse Day Monday, 7am - 9pm

Mariachis De Jalisco Mexican Food Restaurant
2803 Williams Dr #110, (512) 868-5622
Mexican, Georgetown
Eclipse Day Monday, 6am - 10pm

Monument Cafe
500 S Austin Ave, (512) 930-9586
American, Georgetown
Eclipse Day Monday, 7am - 3pm

Tortilleria Y Taqueria San Pedro Limon
905 N Austin Ave, (512) 948-7537
Mexican, Georgetown
Eclipse Day Monday, 7am - 4pm

Tony And Luigi's
1201 S Church St, (512) 864-2687
Italian, Georgetown
Eclipse Day Monday, 11am - 9pm

Wildfire
812 S Austin Ave, (512) 869-3473
Steak, Georgetown
Eclipse Day Monday, 11am - 9pm

Lampasas, TX
Totality Starts at: 1:35pm CDT
Totality Lasts for: 4:24
Totality Ends at: 1:39pm CDT
NWS Office: www.weather.gov/fwd
www.lampasas.org

Alfredo's Mexican Restaurant
2202 US-281, (512) 556-4447
Mexican, Lampasas
Eclipse Day Monday, 11am - 8:30pm

Bella Italia
101-199 W Ave B, (512) 564-5202
Italian, Lampasas
Eclipse Day Monday, 11am - 9pm

Country Kitchen and Bakery
307 N Key Ave, (512) 556-6152
American, Lampasas
Eclipse Day Monday, 7am - 2pm

El Rodeo Mexican Restaurant
609 Central Texas Expy, (512) 556-8757
Mexican, Lampasas
Eclipse Day Monday, 6am - 10pm

Memo's Mexican Restaurant & Bar
407 Central Texas Expy, (512) 564-5166
Mexican, Lampasas
Eclipse Day Monday, 11am - 10pm

Storm's Drive-In Lampasas
201 N Key Ave, (512) 556-6269
American, Lampasas
Eclipse Day Monday, 6:30am - 10pm

Youngs BBQ
504 N Key Ave, (512) 564-1220
BBQ, Lampasas
Eclipse Day Monday, 11am - 6pm

Killeen, TX
Totality Starts at: 1:36pm CDT
Totality Lasts for: 4:17
Totality Ends at: 1:40pm CDT
NWS Office: www.weather.gov/fwd
www.killeentexas.gov

Acropolis
360 E Central Texas Expy # 206, (254) 213-9859
Greek, Harker Heights
Eclipse Day Monday, 11am - 9pm

Arepitas (Harker Heights)
440 E Central Texas Expy Ste 101, (254) 220-4534
Venezuelan, Harker Heights
Eclipse Day Monday, 11am - 9pm

Ban Mai Thai Restaurant
360 E Central Texas Expy, (254) 415-7088
Thai, Harker Heights
Eclipse Day Monday, 11am - 3 pm, 4:30–8pm

Elmore's Fish & Wings
2711 E Veterans Memorial Blvd, (254) 628-8899
Seafood, Killeen
Eclipse Day Monday, 11am - 8pm

Galaxy B&G
104 W Veterans Memorial Blvd, (254) 213-9888
American, Killeen
Eclipse Day Monday, 8am - 10pm

Jamaica Nyammingz Restaurant
1914 E Veterans Memorial Blvd, (254) 432-6968
Jamaican, Killeen
Eclipse Day Monday, 9am - 7pm

Koba Woo Restaurant
332 E Avenue D, (254) 526-3065
Korean, Killeen
Eclipse Day Monday, 11am - 8pm

La Garita Restaurant
305 W Rancier Ave, (254) 213-1513
Puerto Rican, Killeen
Eclipse Day Monday, 11am - 7pm

Los Burritos
2006 W S Young Dr STE 34, (254) 526-8226
Mexican, Killeen
Eclipse Day Monday, 8am - 9pm

New Oriental Restaurant
105 Cox Dr # A Harker Heights, (254) 699-0466
Korean, Harker Heights
Eclipse Day Monday, 11am - 7pm

Simply Good Burgers
301 East Business 190, (254) 577-5900
American, Copperas Cove
Eclipse Day Monday, 11am - 9pm

Taqueria Express Mexican Food
12492 U.S. Hwy 190, (512) 932-2488
Mexican, Kempner
Eclipse Day Monday, 6am - 2pm

Tex-Rican Restaurant
1026 S Fort Hood St, (254) 213-2776
Puerto Rican, Killeen
Eclipse Day Monday, 10:30am - 3pm

Thai Express
4302 E Rancier Ave B, (254) 634-8424
Thai, Killeen
Eclipse Day Monday, 10:30am - 7pm

Toa's Ohana Hawaiian Restaurant
308 E Avenue D, (254) 238-8485
Hawaiian, Copperas Cove
Eclipse Day Monday, 10:30am - 7:30pm

Never gaze at the sun without eye protection. Only remove eye protection during 100% totality.

Temple, TX
Totality Starts at: 1:37pm CDT
Totality Lasts for: 3:49
Totality Ends at: 1:40pm CDT
NWS Office: www.weather.gov/fwd
www.templetx.gov

Bird Creek Burger Co.
6 S Main St, (254) 598-3158
Amercian, Temple
Eclipse Day Monday, 10am - 3pm

Ernie's Fried Chicken
805 S Main St Belton, (254) 939-8032
Chicken, Temple
Eclipse Day Monday, 6am - 7pm

Green's Sausage House
16483 TX-53, (254) 985-2331
American, Temple
Eclipse Day Monday, 7am - 6pm

Hat Creek Burger Company
99 Old Waco Rd, (254) 231-3164
American, Temple
Eclipse Day Monday, 8am - 9pm

Italiano's Pizza, Pasta, & Subs
1316 S 31st St, (254) 791-2000
Pizza, Temple
Eclipse Day Monday, 11am - 9pm

La Dalat Vietnamese Cuisine
17 E Avenue B, (254) 655-7337
Vietnamese, Temple
Eclipse Day Monday, 10:30am - 3 pm, 4:30–9pm

Miller's Smokehouse
300 E Central Ave Belton, (254) 939-5500
BBQ, Temple
Eclipse Day Monday, 7:30 - 3pm

Mee Mee's Authentic Thai Cuisine
3550 S General Bruce Dr, (254) 231-3447
Thai, Temple
Eclipse Day Monday, 11am - 9pm

Megg's Cafe
1749 Everton Dr, (254) 771-3800
American, Temple
Eclipse Day Monday, 7:30am - 3pm

Mosaic Grill
2608 N Main St A, Belton, (254) 613-5240
American and Mediterranean, Belton
Eclipse Day Monday, 11am - 9pm

Old Jody's Restaurant
1219 S 1st St A, (254) 778-9954
American, Temple
Eclipse Day Monday, 10am - 9pm

Pignetti's
14 S 2nd St, (254) 778-1269
Italian, Temple
Eclipse Day Monday, 11am - 9pm

Seoul Garden Temple
7349 Honeysuckle #150, (254) 773-8413
Korean, Temple
Eclipse Day Monday, 11am - 9pm

Sun's Kitchen
1610 S 31st St #100, (254) 231-3304
Asian, Temple
Eclipse Day Monday, 11am - 8pm

Tamarindos
2510 S 5th St, (254) 421-7223
Ice Cream and Snacks, Temple
Eclipse Day Monday, 10am - 9pm

Tapa Tapas
301 N General Bruce Dr, (254) 295-0925
Mexican, Temple
Eclipse Day Monday, 6am - 8pm

Thai Cafe - Thai | Seafood | Oysters
109 W Central Ave, (254) 598-2799
Thai, Temple
Eclipse Day Monday, 11am - 8pm

Gatesville, TX
Totality Starts at: 1:36pm CDT
Totality Lasts for: 4:24
Totality Ends at: 1:41pm CDT
NWS Office: www.weather.gov/fwd
www.gatesvilletx.com

El Tapatio Restaurant #2
1509 E Main St, (254) 865-8039
Mexican, Gatesville
Eclipse Day Monday, 6am - 10pm

J & M's Hill Country Bar-B-Q
2601 E Main St #2629, (254) 865-8706
BBQ, Gatesville
Eclipse Day Monday, 11am - 9pm

Junction on Route 36
1216 TX-36 N, (254) 865-0351
American, Gatesville
Eclipse Day Monday, 8am - 10pm

La Hacienda Mexican Grill
2558 E Main St, (254) 865-5820
Mexican, Gatesville
Eclipse Day Monday, 6am - 9pm

New Baytown Seafood Express
2402 TX-36, (254) 865-0350
Seafood, Gatesville
Eclipse Day Monday, 11am - 9pm

New Rodeo Mexican Grill
309 Highway 36 Byp S, (254) 248-0059
Mexican, Gatesville
Eclipse Day Monday, 6am - 10pm

Park Street Burgers
1602 E Main St suite b, (254) 865-6061
American, Gatesville
Eclipse Day Monday, 10:30am - 2pm

Ranchers Steak House & Grill
107 TX-36 N, (254) 223-2640
Steak, Gatesville
Eclipse Day Monday, 7am - 9pm

Studebakers Pizza
2701 E Main St, (254) 865-8389
Pizza, Gatesville
Eclipse Day Monday, 11am - 9pm

Taco Damian's
4306 s, 4306 TX-36, (254) 326-1501
Mexican, Gatesville
Eclipse Day Monday, 8am - 3pm

Waco, TX
Totality Starts at: 1:37pm CDT
Totality Lasts for: 4:15
Totality Ends at: 1:42pm CDT
NWS Office: www.weather.gov/fwd
www.waco-texas.com

Billy Bob's Burgers Bar & Grill
300 S 2nd St Suite 102, (254) 424-9262
American, Waco
Eclipse Day Monday, 11am - 9pm

Cafe Homestead
388 Halbert Ln, (254) 754-9604
American, Waco
Eclipse Day Monday, 11am - 3pm

Christi's Hamburgers
3101 Beale St Ste A, (254) 799-6480
American, Waco
Eclipse Day Monday, 10:45am - 5:30pm

Dubl-R Old Fashioned Hamburgers
1810 Herring Ave, (254) 753-1603
American, Waco
Eclipse Day Monday, 10am - 4:30pm

El Conquistador
4508 W Waco Dr, (254) 772-4596
Tex-Mex, Waco
Eclipse Day Monday, 11am - 2 pm, 5–10pm

George's Restaurant Bar & Catering
1925 Speight Ave, (254) 753-1421
American, Waco
Eclipse Day Monday, 6:30am - 12 AM

La Familia Restaurant
1111 La Salle Ave, (254) 754-1115
Mexican, Waco
Eclipse Day Monday, 7am - 2pm

Magnolia Table
2132 S Valley Mills Dr, (254) 265-6859
American, Waco
Eclipse Day Monday, 7am - 3pm

Milo
1020 Franklin Ave, (254) 275-6670
American, Waco
Eclipse Day Monday, 11am - 3 pm, 5–9pm

Never gaze at the sun without eye protection. Only remove eye protection during 100% totality.

Revival Eastside Eatery
704 Elm Ave, (254) 339-1401
American, Waco
Eclipse Day Monday, 11am - 3pm

Schmaltz's Sandwich Shoppe
105 S 5th St #2102, (254) 753-2332
Sandwich, Waco
Eclipse Day Monday, 10am - 3pm

Shaan Southern Eatery
1824 W Waco Dr, (254) 754-9926
Seafood, Waco
Eclipse Day Monday, 11am - 8pm

Slow Rise Slice House
7608 Woodway Dr, (254) 235-0785
Pizza, Waco
Eclipse Day Monday, 11am - 10pm

Taqueria Las Trancas
4021 Bellmead Dr, (254) 349-8530
Mexican, Waco
Eclipse Day Monday, 10am - 12 AM

Tejun The Texas Cajun- Robinson
711 Robinson Dr Robinson, (254) 235-3100
Seafood, Waco
Eclipse Day Monday, 11am - 9pm

Tom's Burgers
6818 Sanger Ave, (254) 751-0025
American, Waco
Eclipse Day Monday, 10am - 8pm

Uncle Dan's BBQ & Ribhouse
1001 Lake Air Dr, (254) 772-3532
BBQ, Waco
Eclipse Day Monday, 10:30am - 8pm

Hillsboro, TX
Totality Starts at: 1:38pm CDT
Totality Lasts for: 4:22
Totality Ends at: 1:43pm CDT
NWS Office: www.weather.gov/fwd
www.hillsborotx.org

Branded Burger Co.
100 W Elm St, (972) 775-2202
American, Hillsboro
Eclipse Day Monday, 11am - 8pm

Cilantro's Mexican Restaurant
113 N Waco St, (254) 582-9433
Mexican, Hillsboro
Eclipse Day Monday, 6am - 2pm

El Zarape Mexican Restaurant
106 E Walnut St, (254) 580-9044
Mexican, Hillsboro
Eclipse Day Monday, 7am - 8pm

Milano's Pizza Hillsboro
57 W Franklin St, (254) 283-5245
Pizza, Hillsboro
Eclipse Day Monday, 11am - 9pm

R&K Cafe II
103 N Waco St, (254) 283-5240
American, Hillsboro
Eclipse Day Monday, 6am - 8pm

Williams Drive Inn
328 S Waco St, (254) 582-5327
American, Hillsboro
Eclipse Day Monday, 10am - 8pm

Cleburne, TX
Totality Starts at: 1:39pm CDT
Totality Lasts for: 3:42
Totality Ends at: 1:42pm CDT
NWS Office: www.weather.gov/fwd
www.cleburne.net

Carmelita's Pupuseria & More
1101 Granbury St, (817) 774-2111
Latin, Cleburne
Eclipse Day Monday, 8am - 8pm

Grumps Burgers
1704 N Main St, (817) 774-2874
American, Cleburne
Eclipse Day Monday, 11am - 9pm

Los Vaqueros Mexican Fast Food
280 S Colonial Dr, (817) 526-5182
Mexican, Cleburne
Eclipse Day Monday, 7am - 9:30pm

Morris Neal's Handy Hamburgers
200 S Mill St, (817) 556-6464
American, Cleburne
Eclipse Day Monday, 9:30am - 4pm

Paleteria La Flor De Michoacan
1214 E Henderson St, (682) 317-1465
Mexican, Cleburne
Eclipse Day Monday, 10:30am - 9pm

Thai Garden Cafe LLC
504 S Main St, (817) 645-3666
Thai, Cleburne
Eclipse Day Monday, 11am - 2:30pm

The Garden of Eating Bistro
205 S Main St
American, Cleburne
Eclipse Day Monday, 11am - 2:30pm

Corsicana, TX
Totality Starts at: 1:40pm CDT
Totality Lasts for: 4:10
Totality Ends at: 1:44pm CDT
NWS Office: www.weather.gov/fwd
www.cityofcorsicana.com

Across The Street Diner
125 N Beaton St, (903) 874-9111
American, Corsicana
Eclipse Day Monday, 7am - 4pm

China One
3201 W 7th Ave #105, (903) 872-3888
Chinese, Corsicana
Eclipse Day Monday, 11am - 9pm

Cocina Las Tres Marias
225 N Commerce St, (903) 602-5015
Mexican, Corsicana
Eclipse Day Monday, 6am - 9pm

Corsicana Garden
111 Northwood Blvd, (903) 872-5028
Chinese, Corsicana
Eclipse Day Monday, 11am - 8pm

Ellinya Italian Restaurant
1607 W 7th Ave, (903) 602-5180
Italian, Corsicana
Eclipse Day Monday, 11am - 9pm

El Mexicano Grill
1427 W 7th Ave, (903) 872-8989
Mexican, Corsicana
Eclipse Day Monday, 8am - 9pm

Govea's Mexican Restaurant
316 S 20th St, (903) 229-4906
Mexican, Corsicana
Eclipse Day Monday, 8am - 3pm

Italian Village
510 N 24th St, (903) 874-6804
Italian, Corsicana
Eclipse Day Monday, 11am - 8:30pm

La Cabaña Restaurant
809 N 24th St, (903) 602-5196
Mexican, Corsicana
Eclipse Day Monday, 10am - 8pm

La Pradera Mexican Restaurant
1401 W 7th Ave, (903) 641-0077
Mexican, Corsicana
Eclipse Day Monday, 11am - 9pm

Smokin' Guns BBQ
2728 W 7th Ave, (903) 519-2284
BBQ, Corsicana
Eclipse Day Monday, 11am - 5:30pm

Taqueria Las Comadres
309 S 18th St, (903) 257-3223
Mexican, Corsicana
Eclipse Day Monday, 10am - 7:45pm

Tortilleria Matehuala y Restaurante
105 W 7th Ave #700, (903) 875-1755
Mexican, Corsicana
Eclipse Day Monday, 8am - 3pm

Tucker Town
4095 US-287, (903) 851-3012
BBQ, Corsicana
Eclipse Day Monday, 10:30am - 8pm

Ennis, TX
Totality Starts at: 1:39pm CDT
Totality Lasts for: 4:22
Totality Ends at: 1:44pm CDT
NWS Office: www.weather.gov/fwd
www.ennistx.gov

Alma Smokehouse BBQ Alma
106 SW Interstate 45, (972) 875-2669
BBQ, Alma
Eclipse Day Monday, 8am - 8pm

Never gaze at the sun without eye protection. Only remove eye protection during 100% totality.

Bailey's Cafe
508 S Kaufman St, (972) 875-1556
American, Ennis
Eclipse Day Monday, 11am - 3pm

Birrieria Aguinaga
905 E Ennis Ave, (972) 876-4232
Mexican, Ennis
Eclipse Day Monday, 8am - 10pm

Bubba's BBQ & Steakhouse
210 I-45, (972) 875-0036
BBQ, Ennis
Eclipse Day Monday, 11am - 8pm

Ennis Family Restaurant
2709 S Kaufman St # A, (972) 875-7200
American, Ennis
Eclipse Day Monday, 7am - 2pm

From Scratch Daniel's Kitchen
218 W Ennis Ave, (469) 456-0067
American, Ennis
Eclipse Day Monday, 7am - 10pm

Grand Ennis Buffet
201 S I-45 #a, (972) 875-9988
Buffet, Ennis
Eclipse Day Monday, 11am - 9pm

Hilda's Kitchen Mexican Food
509 E Ennis Ave, (972) 875-2370
Mexican, Ennis
Eclipse Day Monday, 5:30am - 2pm

Limerick's Sandwich Junction
213 W Ennis Ave Ste 100, (903) 405-1995
Sandwich, Ennis
Eclipse Day Monday, 11am - 7pm

Nortenos Mexican Food
814 W Ennis Ave, (972) 875-9335
Mexican, Ennis
Eclipse Day Monday, 10am - 3pm

Reina's Taqueria
107 E Brown St, (469) 881-1396
Mexican, Ennis
Eclipse Day Monday, 7am - 9pm

Taqueria Dona Mari
1405 S Kaufman St, (972) 875-9166
Mexican, Ennis
Eclipse Day Monday, 7am - 4pm

Waxahachie, TX
Totality Starts at: 1:39pm CDT
Totality Lasts for: 4:17
Totality Ends at: 1:44pm CDT
NWS Office: www.weather.gov/fwd
www.waxahachie.com

Butter & Grace
1585 N Hwy 77, (214) 980-1679
American, Waxahachie
Eclipse Day Monday, 6am - 3pm

Campuzano's Fine Mexican Food
2167 N Hwy 77, (972) 938-0047
Mexican, Waxahachie
Eclipse Day Monday, 11am - 9pm

Country Cafe
217 US-77, (972) 923-0214
American, Waxahachie
Eclipse Day Monday, 6:30am - 2pm

Country Cafe
217 US-77, (972) 923-0214
American, Waxahachie
Eclipse Day Monday, 6:30am - 2pm

El Mexicano Grill
114 E Franklin St, (972) 937-1191
Mexican, Waxahachie
Eclipse Day Monday, 9am - 12 AM

El Trebol Taqueria
1204 Ferris Ave F, (972) 937-5373
Mexican, Waxahachie
Eclipse Day Monday, 8am - 7:30pm

Farm Luck Soda Fountain & Dry Goods
109 W Franklin St #119, (214) 903-8021
American, Waxahachie
Eclipse Day Monday, 11am - 9pm

Hibachio
503 N Hwy 77, (214) 903-8048
Japanese, Waxahachie
Eclipse Day Monday, 11am - 9pm

La Norteñita Grill
1102 Ferris Ave, (214) 463-5503
Mexican, Waxahachie
Eclipse Day Monday, 10am - 10pm

Oma's Jiffy Burger
403 Water St, (972) 937-9190
American, Waxahachie
Eclipse Day Monday, 7am - 3pm

Pop's Burger Stand
107 S Monroe St, (972) 923-8922
American, Waxahachie
Eclipse Day Monday, 11am - 8pm

Taco Suave
121 N Hwy 77, (214) 980-1300
Mexican, Waxahachie
Eclipse Day Monday, 8am - 9pm

Tacos Facio y Guisado
212 W Jefferson St Suite 2, (469) 553-9857
Mexican, Waxahachie
Eclipse Day Monday, 7:30am - 3pm

TaMolly's of Waxahachie
1735 N Hwy 77, (972) 937-7772
Mexican, Waxahachie
Eclipse Day Monday, 11am - 8pm

Tomatoes Mexican & Italian
619 Ferris Ave, (972) 937-7931
Mexican & Italian, Waxahachie
Eclipse Day Monday, 11am - 9pm

Two Amigos Taqueria
241 S Monroe St, (972) 923-3305
Mexican, Waxahachie
Eclipse Day Monday, 7am - 9pm

Athens, TX
Totality Starts at: 1:41pm CDT
Totality Lasts for: 3:17
Totality Ends at: 1:44pm CDT
NWS Office: www.weather.gov/fwd
www.athenstx.gov

Bensons Eats & Treats
319 S Palestine St, (903) 675-1414
American, Athens
Eclipse Day Monday, 10am - 10:30pm

Cherry Laurel Bakery, Cafe & Catering
305 S Prairieville St, (903) 677-5599
American, Athens
Eclipse Day Monday, 8am - 3pm

Danny's Smokehouse Bar-B-Q
850 E Corsicana St, (903) 675-5238
BBQ, Athens
Eclipse Day Monday, 10am - 8:30pm

Mariscos El Rincon Bar & Grill
300 N Pinkerton St, (903) 264-1010
Seafood, Athens
Eclipse Day Monday, 10am - 9pm

Neza Mexican Cuisine
416 S Palestine St c, (903) 292-5279
Mexican, Athens
Eclipse Day Monday, 11am - 9pm

Ochoa's Mexican Restaurant
310 US-175 BUS, (903) 677-1316
Mexican, Athens
Eclipse Day Monday, 11am - 8pm

Pitt Grill Restaurant
520 W Corsicana St, (903) 670-3207
American, Athens
Eclipse Day Monday, Open 24 hours

Restaurant y Taqueria Jalisco's
506 W Corsicana St, (903) 677-6169
Mexican, Athens
Eclipse Day Monday, 7am - 9pm

Twisted Root Burger Co.
1012 E Tyler St Bldg A, (430) 340-5230
American, Athens
Eclipse Day Monday, 11am - 9pm

Yamato Sushi & Steak House
612 W Corsicana St, (903) 292-1223
Sushi & Steak, Athens
Eclipse Day Monday, 11am - 9:30pm

Tyler, TX
Totality Starts at: 1:43pm CDT
Totality Lasts for: 2:13
Totality Ends at: 1:45pm CDT
NWS Office: www.weather.gov/shv
www.cityoftyler.org

Athena Greek & American Family Restaurant
1593 W SW Loop 323, (903) 561-8065
Greek & American, Tyler
Eclipse Day Monday, 11am - 9pm

Never gaze at the sun without eye protection. Only remove eye protection during 100% totality.

Bar BQ Hernandez
501 S Glenwod Blvd, (903) 593-3973
BBQ, Tyler
Eclipse Day Monday, 10am - 5pm

Bruno's | Pizza & Pasta
15770 FM2493, (903) 939-0002
Pizza, Tyler
Eclipse Day Monday, 11am - 9pm

Burger Warehouse
1839 Troup Hwy
American, Tyler
Eclipse Day Monday, 11am - 7pm

Clear Springs Restaurant
6519 S Broadway Ave, (903) 561-0700
Seafood, Tyler
Eclipse Day Monday, 11am - 9pm

Daniel Boone's Grill & Tavern
1920 E SE Loop 323, (903) 595-2228
American, Tyler
Eclipse Day Monday, 11am - 8pm

Jucys Hamburgers
2330 E 5th St, (903) 597-0660
American, Tyler
Eclipse Day Monday, 6am - 10pm

Loggins Restaurant
137 S Glenwood Blvd, (903) 595-5022
American, Tyler
Eclipse Day Monday, 10:30am - 2pm

Mama's Restaurant
2105 E 5th St, (903) 526-7915
American, Tyler
Eclipse Day Monday, 6am - 10pm

Razzoo's Cajun Cafe
7011 S Broadway Ave, (903) 534-2922
Cajun, Tyler
Eclipse Day Monday, 11am - 9pm

Roost
3314 Troup Hwy, (903) 200-1775
Sandwich, Tyler
Eclipse Day Monday, 7am - 7pm

Ruby's Mexican Restaurant
2021 E Gentry Pkwy, (903) 617-6816
Mexican, Tyler
Eclipse Day Monday, 7am - 8pm

Taqueria Flor de Taxco
2522 Shiloh Rd, (469) 565-5954
Mexican, Tyler
Eclipse Day Monday, 7am - 9pm

Taco Shop Tyler Mexican Restaurant
7205 S Broadway Ave Suite 100, (903) 630-1927
Mexican, Tyler
Eclipse Day Monday, 10:30am - 9:30pm

The Diner
7924 S Broadway Ave, (903) 509-3463
Breakfast, Tyler
Eclipse Day Monday, 6am - 2pm

The Potpourri House
3320 Troup Hwy Ste. 300, (903) 592-4171
American, Tyler
Eclipse Day Monday, 10:30am - 3pm

Tiba Grill
211 Shelley Dr, (903) 561-8335
Mediterranean, Tyler
Eclipse Day Monday, 11am - 8pm

Topp's Pizza
3101 Shiloh Rd #131, (903) 258-6111
Pizza, Tyler
Eclipse Day Monday, 11am - 10pm

Yamato
2210 W SW Loop 323, (903) 534-1888
Japanese, Tyler
Eclipse Day Monday, 11am - 10pm

Dallas-Fort Worth
Totality Starts at: 1:40pm CDT
Totality Lasts for: D 3:50, FW 2:35
Totality Ends at: 1:44pm CDT
NWS Office: www.weather.gov/fwd
dallascityhall.com, fortworthtexas.gov

Bells Better Burgers
2610 Peachtree Rd Balch Springs, (972) 286-0320
American, Balch Springs
Eclipse Day Monday, 10am - 9pm

Country Burger
401 S Hampton Rd Dallas, (214) 330-4743
American, Dallas
Eclipse Day Monday, 10:30am - 9:30pm

Dough City Pizza+Burgers
219 TX-342 suite 120 Red Oak, (469) 820-9616
Pizza, Red Oak
Eclipse Day Monday, 11am - 8pm

El Mofongo Restaurant
3701 S Cooper St #135 Arlington, (817) 466-8802
Puerto Rican, Arlington
Eclipse Day Monday, 11am - 8pm

Fattoush Restaurant
2304 W Park Row Dr Suite 25, (682) 321-7650
Mediterranean & Iraqi-style, Pantego
Eclipse Day Monday, 11am - 9pm

Heim Barbecue
1109 W Magnolia Ave, Forth Worth (817) 882-6970
BBQ, Forth Worth
Eclipse Day Monday, 9am - 9pm

Hideout Burgers
1601 S 9th St #500 Midlothian, (972) 775-4949
American, Midlothian
Eclipse Day Monday, 10am - 9pm

Hot Rodz Diner LLC
115 S Main St Crandall, (972) 472-3714
American, Crandall
Eclipse Day Monday, 7am - 8pm

Istanbul Grill
401 Throckmorton St, Fort Worth (817) 885-7326
Mediterranean, Fort Worth
Eclipse Day Monday, 11am - 9pm

Jalapenos Lemon Pepper
400 W Avenue F Midlothian, (972) 775-1008
Tex-Mex, Midlothian
Eclipse Day Monday, 11am - 9pm

Juniors Grill
2406 W Park Row Dr Pantego, (682) 323-8652
American, Pantego
Eclipse Day Monday, 7am - 8pm

Kaufman Burgers
807 S Washington St #2709, (469) 208-5808
American, Kaufman
Eclipse Day Monday, 11am - 8:30pm

Khanh's Vietnamese Restaurant
6901 Mc Cart Ave # 200, (817) 720-3200
Vietnamese, Fort Worth
Eclipse Day Monday, 10am - 9pm

Lai Lai China Restaurant
8323 Lake June Rd Dallas, (214) 398-4101
Chinese, Dallas
Eclipse Day Monday, 11am - 5pm

Lindy's Restaurant
13425 Seagoville Rd Dallas, (972) 557-5659
American, Dallas
Eclipse Day Monday, 7am - 1:45pm

Myrtle's Burgers
4568 E, FM1187 # B Burleson, (817) 561-1836
American, Burleson
Eclipse Day Monday, 11am - 5pm

Norma's Cafe
1123 W Davis St Dallas, (214) 946-4711
American, Dallas
Eclipse Day Monday, 7am - 8pm

Pecan Lodge
2702 Main St, Dallas, (214) 748-8900
BBQ, Dallas
Eclipse Day Monday, 11am - 3pm

Prespa's Italian Restaurant
4720 W Sublett Rd, Arlington, (817) 561-7540
Italian, Arlington
Eclipse Day Monday, 11am - 9pm

Reata Restaurant
310 Houston St, Fort Worth (817) 336-1009
Southwestern, Fort Worth
Eclipse Day Monday, 11am - 2:30 pm, 5–8:30pm

Skillet & Grill Inc
1801 W Division St Arlington, (817) 795-8682
Breakfast, Arlington
Eclipse Day Monday, 5am - 3pm

Sweet Georgia Brown
2840 E Ledbetter Dr Dallas, (214) 375-2020
Soul Food, Dallas
Eclipse Day Monday, 11am - 9pm

Terry Black's Barbecue
3025 Main St, Dallas, (469) 399-0081
BBQ, Dallas
Eclipse Day Monday, 11am - 9:30pm

The Porch
140 S Wilson St Burleson, (817) 426-9900
Amercian, Burleson
Eclipse Day Monday, 6am - 8pm

Never gaze at the sun without eye protection. Only remove eye protection during 100% totality.

Theo's Grill & Bar
107 NW 8th St Grand Prairie, (972) 262-8886
American, Grand Prairie
Eclipse Day Monday, 7am - 3pm

Tootie's Southern Kitchen
S 4606 Botham Jean Blvd Dallas, (214) 421-3545
Soul Food, Dallas
Eclipse Day Monday, 7am - 10pm

Wings Over Seagoville
1701 N Hwy 175 Seagoville, (972) 287-1191
Wings, Seagoville
Eclipse Day Monday, 11am - 10pm

Sulphur Springs, TX
Totality Starts at: 1:42pm CDT
Totality Lasts for: 4:20
Totality Ends at: 1:47pm CDT
NWS Office: www.weather.gov/fwd
www.sulphurspringstx.org

Bodacious Bar-B-Q
1228 S Broadway St, (903) 885-6456
BBQ, Sulphur Springs
Eclipse Day Monday, 10:30am - 9pm

Broadway Buffet
1145 S Broadway St, (903) 885-7959
Buffet, Sulphur Springs
Eclipse Day Monday, 10:30am - 9pm

Chen's Kitchen
472 Shannon Rd W, (903) 335-8195
Chinese, Sulphur Springs
Eclipse Day Monday, 12–8:30pm

Corner Grub House
113 Gilmer St, (903) 919-5058
American, Sulphur Springs
Eclipse Day Monday, 10:30am - 9pm

Don Lalo's, SS
912 Gilmer St, (903) 558-1859
Mexican, Sulphur Springs
Eclipse Day Monday, 9am - 5pm

Flips Burgerland
210 Main St, (903) 919-5031
American, Sulphur Springs
Eclipse Day Monday, 11am - 2:30pm

Metro Diner
105 Industrial Dr W, (903) 885-6446
American, Sulphur Springs
Eclipse Day Monday, Open 24 hours

Pioneer Cafe
325 Jefferson St E, (903) 885-7773
American, Sulphur Springs
Eclipse Day Monday, 6am - 2pm

Redbarn Cafe
1310 Hillcrest Dr N, (903) 885-3332
American, Sulphur Springs
Eclipse Day Monday, 7am - 2pm

Sakura Sushi & Seafood Grill
1408 S Broadway St suit 150, (903) 438-0088
Sushi, Sulphur Springs
Eclipse Day Monday, 11am - 9:30pm

Soulmans - Sulphur Springs BBQ
1201 S Broadway St, (903) 919-3617
BBQ, Sulphur Springs
Eclipse Day Monday, 11am - 9pm

Tradicion Mexicana Restaurant and Tortilla Factory
325 Industrial Dr W, (903) 438-9450
Mexican, Sulphur Springs
Eclipse Day Monday, 9am - 7pm

Paris, TX
Totality Starts at: 1:44pm CDT
Totality Lasts for: 4:00
Totality Ends at: 1:48pm CDT
NWS Office: www.weather.gov/fwd
www.paristexas.gov

Burgerland
1301 N Main St, (903) 739-9443
American, Paris
Eclipse Day Monday, 10:30am - 8pm

Capizzi's Italian Kitchen
2525 Clarksville St, (903) 785-7590
Italian, Paris
Eclipse Day Monday, 11am - 9pm

China Star
1810 Lamar Ave, (903) 785-8872
Chinese, Paris
Eclipse Day Monday, 11am - 2:30 pm, 5–8:30pm

Crawford's Hole In the Wall
202 3rd St NW, (903) 737-9025
American, Paris
Eclipse Day Monday, 10am - 2pm

Dos Marias
303 20th St NE, (903) 737-4557
Mexican, Paris
Eclipse Day Monday, 11am - 9pm

Golden China
3755 Lamar Ave, (903) 739-8860
Chinese, Paris
Eclipse Day Monday, 11am - 8:30pm

High Cotton Kitchen
1260 Clarksville St, (903) 385-3885
American, Paris
Eclipse Day Monday, 7am - 9pm

Jaxx Burgers
10 Clarksville St, (903) 739-2955
American, Paris
Eclipse Day Monday, 11am - 8pm

La Bella Airosa
1335 19th St SW, (903) 783-1181
Mexican, Paris
Eclipse Day Monday, 10:30am - 2pm

Lance's Sunriser
2880 N Main St, (903) 785-9149
American, Paris
Eclipse Day Monday, Open 24 hours

Luna Azul Mexican Restaurant
1603 N Main St, (903) 609-8006
Mexican, Paris
Eclipse Day Monday, 11am - 9pm

Magel's Grill
3805 NE Loop 286, (903) 784-3186
American, Paris
Eclipse Day Monday, 11am - 8pm

McKee's Family Restaurant
1355 N Collegiate Dr, (903) 785-0002
American, Paris
Eclipse Day Monday, 6am - 9pm

Phat Phil's BBQ
1343 S Church St, (903) 782-9545
BBQ, Paris
Eclipse Day Monday, 11am - 4pm

Sandwich Etc.
3060 Lamar Ave, (903) 784-1626
Sandwich, Paris
Eclipse Day Monday, 7:30am - 3pm

Scholl Bros. BBQ
1528 Lamar Ave, (903) 739-8080
BBQ, Paris
Eclipse Day Monday, 11am - 8pm

South Main Cafe
1222 S Main St, (903) 783-9200
American, Paris
Eclipse Day Monday, 6:30am - 3pm

Sr. Submarine & Gyros
1250 Lamar Ave, (903) 401-5233
Sandwich, Paris
Eclipse Day Monday, 10am - 8pm

Taco Delite
1580 Clarksville St, (903) 785-7173
Mexican, Paris
Eclipse Day Monday, 7am - 10pm

Tokyo Express in Paris
3090 Clarksville St, (903) 706-5777
Japanese, Paris
Eclipse Day Monday, 11am - 9pm

Torres Mochas
1995 Lamar Ave, (903) 401-5033
Mexican, Paris
Eclipse Day Monday, 11am - 2 pm, 4:30–8pm

Mt Pleasant, TX
Totality Starts at: 1:44pm CDT
Totality Lasts for: 3:56
Totality Ends at: 1:48pm CDT
NWS Office: www.weather.gov/shv
mpcity.net

Country Cafe Diner
1410 W Ferguson Rd, (903) 717-8644
American, Mt Pleasant
Eclipse Day Monday, 7am - 2:30pm

Herschel's Restaurant
1612 S Jefferson Ave, (903) 572-7801
American, Mt Pleasant
Eclipse Day Monday, 6:30am - 10pm

Never gaze at the sun without eye protection. Only remove eye protection during 100% totality.

JoJack's Smokehouse
2310 N Jefferson Ave, (903) 717-8301
BBQ, Mt Pleasant
Eclipse Day Monday, 11am - 8pm

Lala's Mexican Food
1649 Farm to Market 1001, (903) 575-0229
Mexican, Mt Pleasant
Eclipse Day Monday, 11am - 8pm

Mexico Restaurant
301 W Ferguson Rd, (903) 577-9421
Mexican, Mt Pleasant
Eclipse Day Monday, 10am - 9pm

Mount Pleasant Burgers & Fries
100 E 14th St, (903) 717-8917
American, Mt Pleasant
Eclipse Day Monday, 10:30am - 7pm

Mt Fuji Hibachi Steakhouse
650 S Jefferson Ave, (903) 577-1700
Japanese, Mt Pleasant
Eclipse Day Monday, 11am - 3 pm, 5–9pm

Nardello's
103 N Madison, (903) 380-6200
Pizza, Mt Pleasant
Eclipse Day Monday, 11am - 9pm

Outlaw's Bar-B-Que
100 W Ferguson Rd, (903) 572-7860
BBQ, Mt Pleasant
Eclipse Day Monday, 10:30am - 8pm

Pollo Bueno
315 W Ferguson Rd, (903) 572-3368
Mexican, Mt Pleasant
Eclipse Day Monday, 7am - 10pm

Randy's Burgers
100 N Edwards Ave, (903) 572-2666
American, Mt Pleasant
Eclipse Day Monday, 11am - 7pm

Taqueria Monterrey
721 W 12th St, (903) 575-9447
Mexican, Mt Pleasant
Eclipse Day Monday, 10am - 8pm

Thai Lanna
208 Lakewood Dr, (903) 577-1500
Thai, Mt Pleasant
Eclipse Day Monday, 11am - 9pm

Clarksville, TX
Totality Starts at: 1:44pm CDT
Totality Lasts for: 4:19
Totality Ends at: 1:48pm CDT
NWS Office: www.weather.gov/shv
clarksvilletx.com

Cheyenne BBQ
107 Delaware St, (903) 427-0510
BBQ, Clarksville
Eclipse Day Monday, 11am - 7pm

Chilango's Mexican Restaurant
2000 Main St, (903) 427-0089
Mexican, Clarksville
Eclipse Day Monday, 10:30am - 9pm

Coleman Barbecue
604 N Donoho St, (903) 427-5131
BBQ, Clarksville
Eclipse Day Monday, 11am - 3pm

El Patrón Mexican Grill
2000 Main St, (903) 427-0089
Mexican, Clarksville
Eclipse Day Monday, 10:30am - 9pm

Rio Verde Restaurant
400 E Main St, (903) 428-0030
Mexican, Clarksville
Eclipse Day Monday, 11am - 8pm

Texarkana, TX / AR
Totality Starts at: 1:46pm CDT
Totality Lasts for: 2:35
Totality Ends at: 1:49pm CDT
NWS Office: www.weather.gov/shv
www.ci.texarkana.tx.us

Colima's
3505 Summerhill Rd TX, (903) 792-1166
Mexican, Texarkana
Eclipse Day Monday, 8:30am - 7:30pm

Gusanos Chicago Style Pizzeria
2820 Richmond Rd TX, (903) 792-8646
Pizza, Texarkana
Eclipse Day Monday, 11am - 9pm

Ironwood Grill
4312 Galleria Oaks Dr Tx, (903) 223-4644
American, Texarkana
Eclipse Day Monday, 11am - 9pm

Johnny B's of Texarkana
224 E 5th St AR, (870) 330-9471
American, Texarkana
Eclipse Day Monday, 6am - 3pm

Loca Luna Mexican Grill
5200 W 7th St TX, (903) 306-0096
Mexican, Texarkana
Eclipse Day Monday, 11am - 9:30pm

Lunch Box
503 S Robison Rd TX, (903) 832-3874
American, Texarkana
Eclipse Day Monday, 11am - 2pm

M & M Burgers & Etc
220 East St AR, (870) 774-6828
American, Texarkana
Eclipse Day Monday, 10am - 6pm

Naaman's BBQ
5200 N State Line Ave AR, (903) 826-2827
BBQ, Texarkana
Eclipse Day Monday, 10:30am - 9pm

Old Tyme Burger Shoppe
1205 Arkansas Blvd AR, (870) 772-5775
American, Texarkana
Eclipse Day Monday, 7am - 7pm

Reggie's Burgers Dogs & Fries
3200 N State Line Ave AR, (870) 330-4447
American, Texarkana
Eclipse Day Monday, 10:30am - 3pm

Silver Star Smokehouse
5205 W Park Blvd TX, (903) 306-0778
BBQ, Texarkana
Eclipse Day Monday, 11am - 9pm

Sue N' Carol's Kitchen
938 N State Line Ave AR, (870) 774-0859
American, Texarkana
Eclipse Day Monday, 7am - 3pm

The Dugout
3801 E 9th St AR, (870) 330-4109
American, Texarkana
Eclipse Day Monday, 10:30am - 8pm

The Pully Bone
240 E New Boston Rd Nash TX, (903) 306-1340
American, Texarkana
Eclipse Day Monday, 6am - 6pm

Three Chicks Feed, Seed & Cafe
4045 Genoa Rd AR, (870) 773-5633
American, Texarkana
Eclipse Day Monday, 8am - 5:30pm

TLC Burgers & Fries
201 E Broad St AR, (870) 773-9316
American, Texarkana
Eclipse Day Monday, 10:30am - 3:30pm

Wayne's Family Restaurant
2219 S Lake Dr TX, (903) 792-3561
American, Texarkana
Eclipse Day Monday, 10am - 8pm

Oklahoma

Idabel, OK
Totality Starts at: 1:45pm CDT
Totality Lasts for: 4:18
Totality Ends at: 1:49pm CDT
NWS Office: www.weather.gov/shv
City Website: www.idabel-ok.gov

David Beard's Catfish Seafood And Steak House,
2501 SE Washington St, (580) 286-3387,
Seafood, Idabel
Eclipse Day Monday, 11am - 9pm

Delicias Mexicanas,
9 NW Martin Luther King Dr, (903) 905-0607,
Mexican, Idabel
Eclipse Day Monday, 11am - 8pm

Gemini Coffee Shop,
421 S Central Ave, (580) 286-2900,
American, Idabel
Eclipse Day Monday, 7am - 2pm

Never gaze at the sun without eye protection. Only remove eye protection during 100% totality.

Jake's Brickhouse Grill,
14 SW Main St, (580) 262-6064,
American, Idabel
Eclipse Day Monday, 11am - 9pm

Lin Cuisine,
1810 SE Washington St, (580) 208-2686,
Chinese, Idabel
Eclipse Day Monday, 11am - 9pm

Milano's Pizza,
2106 SE Washington St, (580) 286-6400,
Pizza, Idabel
Eclipse Day Monday, 10:30am - 11pm

Taco Alley,
1103 SE Washington St, (903) 824-3821,
Mexican, Idabel
Eclipse Day Monday, 10:30am - 8pm

Broken Bow, OK
Totality Starts at: 1:45pm CDT
Totality Lasts for: 4:16
Totality Ends at: 1:50pm CDT
NWS Office: www.weather.gov/shv
City Website: cityofbrokenbow.com

Mexico Lindo,
2405 S Park Dr, (580) 584-3287,
Mexican, Broken Bow
Eclipse Day Monday, 10:30am - 9pm

Mi Ranchito Tex Mex Restaurant,
406 S Park Dr, (580) 584-2003,
Mexican, Broken Bow
Eclipse Day Monday, 10:30am - 9pm

Oaks Steak House & Gifts,
2204 S Park Dr, (580) 584-5266,
American, Broken Bow
Eclipse Day Monday, 11am - 8:30pm

Papa Poblanos Broken Bow,
304 N Park Dr, (580) 584-9495,
Mexican, Broken Bow
Eclipse Day Monday, 11am - 9pm

Smith Good Eats,
204 S Park Dr, (580) 584-3988,
American, Broken Bow
Eclipse Day Monday, 7am - 2pm

Sunrise Buffet,
705 S Park Dr, (580) 584-9328,
Chinese, Broken Bow
Eclipse Day Monday, 11am - 9:30pm

Featured Parks and Attractions

Beavers Bend State Park in Oklahoma, is situated along the shores of Broken Bow Lake and offers water activities like swimming, fishing, and canoeing. Hiking enthusiasts can explore miles of trails through pine and hardwood forests. Other attractions include horseback riding, mini-golf, and the Forest Heritage Center Museum. Cabins and campsites are available for overnight stays. On April 4, 2024, day turns to night at this location for over four minutes, so be ready to marvel at the stunning solar eclipse while you enjoy the park.

The Dallas Arboretum & Botanical Garden is 66-acres of stunning gardens along White Rock Lake. Visitors can enjoy seasonal outdoor festivals, concerts, art shows, and exquisite dining options at Restaurant DeGolyer or the Lula Mae Slaughter Dining Terrace. Limited free tickets can be obtained at City of Dallas recreation centers on a first-come basis for individuals. On eclipse day you'll experience nearly 4 minutes of totality near this interesting tourist destination.

Dinosaur Valley State Park, located in Glen Rose, Texas, is a unique destination that transports visitors back to the prehistoric era. The park is renowned for its well-preserved dinosaur tracks imprinted in the bed of the Paluxy River. The park also offers a variety of outdoor activities such as hiking, mountain biking, bird watching, and camping against the backdrop of scenic landscapes. The 2024 eclipse will darken the sky over the dinosaur tracks for three minutes.

Enchanted Rock State Natural Area, in the Texas Hill Country, is a magnet for outdoor enthusiasts and geology aficionados. Its centerpiece is a massive pink granite dome, the Enchanted Rock, rising 425 feet above the surrounding landscape, and considered sacred by local Native American tribes. Visitors can hike to the summit for panoramic views of the area, or explore the park's many trails. The sun's temporary retreat at this park lasts for 4:24, nearly the maximum duration for this total solar eclipse.

Garner State Park, located in Concan, Texas, is an outdoor lover's paradise with its picturesque views, crystal clear waters, and abundant wildlife. With over 11 miles of scenic hiking trails that range from easy to challenging, visitors can enjoy breathtaking views of the Texas Hill Country. Prepare to be dazzled by nearly four and half minutes of the total solar eclipse while enjoying this amazing state park.

The Perot Museum of Nature and Science in Dallas, provides many exhibits and experiences. Engage your imagination with exhibits ranging from the Rose Hall of Birds, where you can explore the connection between prehistoric dinosaurs and modern-day birds, to the Texas Instruments Engineering and Innovation Hall, where you can take on challenges in construction, 3D animation, and robotics. Be sure to step outside before 1:40pm, because the total solar eclipse will stun onlookers for nearly four minutes at this location.

The parks and attractions section in this book are organized by cities, which are highlighted in bold. Under each city, you will find a curated list of locations presented in alphabetical order. These locations may be within the city or situated within a 30-50 mile radius, encompassing the surrounding region. This method of organization allows for easy navigation and planning, ensuring you can make the most of your solar eclipse experience.

Never gaze at the sun without eye protection. Only remove eye protection during 100% totality.

Parks and Attractions

Eagle Pass, TX
Totality Starts at: 1:27pm CDT
Totality Lasts for: 4:22
Totality Ends at: 1:27pm CDT
NWS Office: www.weather.gov/ewx
City Website: www.eaglepasstx.us

4 Amigos Ranch
9061 State Loop 480
Eagle Pass, TX 78852
www.4amigosranch.com

Eagle Pass Lake
Maverick County Lake Jogging Trail
Eagle Pass, TX 78852
www.eaglepasschamber.com

Jando Guedea Park
413 Nueces St
Eagle Pass, TX 78852
eaglepasstx.com/jando-guedea-park/

Kickapoo Lucky Eagle Casino RV Park Campground
192 Quail Valley Rd
Eagle Pass, TX 78852
luckyeagletexas.com/rv-park/

Pasquale DeBona Park
2092 Flowers Dr
Eagle Pass, TX 78852
Totality Lasts for: 4:22

San Juan Park
600 Madison St
Eagle Pass, TX 78852
www.eaglepasstx.us

Del Rio, TX
Totality Starts at: 1:28pm CDT
Totality Lasts for: 3:24
Totality Ends at: 1:31pm CDT
NWS Office: www.weather.gov/ewx
City Website: www.cityofdelrio.com

277 North Campground NPS
Del Rio, TX 78840
www.nps.gov/amis/
Totality Lasts for: 2:55

Buzzard Roost RV Barn
4411 Veterans Blvd
Del Rio, TX 78840
(830) 774-5151

Amistad National Recreation Area
4121 Veterans Blvd, Del Rio, TX 78840
www.nps.gov/amis/
Totality Lasts for: 2:03

Governor's Landing Campground NPS
US-90, Del Rio, TX 78840
www.nps.gov/amis/
Totality Lasts for: 2:35

Laughlin Heritage Foundation Museum
309 S Main St
Del Rio, (830) 719-9380
www.laughlinheritagefoundationmuseum.org

San Pedro Campground NPS
Yellowstone Dr, Del Rio, TX 78840
www.nps.gov/amis/
Totality Lasts for: 2:54

Whitehead Memorial Museum
1308 S Main St, Del Rio
(830) 774-7568
whiteheadmuseum.org

Moore Park
100 Swift Street
Del Rio, TX 78840
www.cityofdelrio.com

Brackettville, TX
Totality Starts at: 1:28pm CDT
Totality Lasts for: 4:15
Totality Ends at: 1:33pm CDT
NWS Office: www.weather.gov/ewx
www.thecityofbrackettville.com

Kickapoo Cavern State Park
20939 RTM Rd 674 N, Brackettville, TX
tpwd.state.tx.us
Totality Lasts for: 3:51

Fort Clark Museum
152 McClernand Rd
Brackettville, (830) 563-2493
www.fortclark.com

Carrizo Springs, TX
Totality Starts at: 1:29pm CDT
Totality Lasts for: 2:51
Totality Ends at: 1:31pm CDT
NWS Office: www.weather.gov/ewx
City Website: www.cityofcarrizo.org

Brush Country Oasis RV Park
201 N 23rd St
Carrizo Springs, TX 78834
brushcountryoasisrvpark.com

Carrizo Springs RV Park
2211 N 1st St
Carrizo Springs, TX 78834
(830) 876-3059

SB-RV Park
205 Fig Ave
Carrizo Springs, TX 78834
(830) 219-9218

Texas Olive Ranch
2488 FM 1557, Carrizo Springs, TX 78834
Olive Oil Producer
texasoliveranch.com

Veterans Park
405 E Alamo St
Carrizo Springs, TX 78834
www.cityofcarrizo.org

Crystal City, TX
Totality Starts at: 1:29pm CDT
Totality Lasts for: 3:16
Totality Ends at: 1:32pm CDT
NWS Office: www.weather.gov/ewx
City Website: www.crystalcitytx.org

Cross S RV Park
3780 US-83
Crystal City, TX 78839
www.crosssrvpark.com

Juan Garcia Park
1531 US-83
Crystal City, TX 78839
crystalcitytexas.com/juan-garcia-park/

Popeye Statue, Crystal City Texas
Spinach Capital of the World
400 W Zavala St, Crystal City, TX 78839
texashillcountry.com

Uvalde, TX
Totality Starts at: 1:29pm CDT
Totality Lasts for: 4:16
Totality Ends at: 1:33pm CDT
NWS Office: www.weather.gov/ewx
City Website: www.uvaldetx.gov

Cook's Slough Sanctuary
Uvalde, TX 78801
Bird Watching Area
texastimetravel.com

Fort Inge
399 Co Rd 375, Uvalde, TX 78801
1849 Federal Fort
www.legendsofamerica.com/fort-inge-texas/

Friends of John Garner Museum
200 E Nopal St # 211
Uvalde, (830) 278-5018
briscoecenter.org/briscoe-garner-museum/

Never gaze at the sun without eye protection. Only remove eye protection during 100% totality.

Garner State Park
234 RR 1050, Concan, TX 78838
tpwd.state.tx.us
Totality Lasts for: 4:26

Quail Springs RV Park Community
2727 E Main St
Uvalde, TX 78801
quailspringsrvpark.com

Rv Park (MTZ Rv Park)
1919 Crystal City Hwy
Uvalde, TX 78801
www.mtzrvpark.com

Uvalde National Fish Hatchery
754 County Rd 203, Uvalde, (830) 278-2419
2,100 miles of refreshing trails
fws.gov/fish-hatchery/uvalde

Weisman Museum of Southwest Texas
301 W Main St
Uvalde, TX 78801
Totality Lasts for: 4:16

Leakey, TX
Totality Starts at: 1:30pm CDT
Totality Lasts for: 4:25
Totality Ends at: 1:34pm CDT
NWS Office: www.weather.gov/ewx
City Website: www.co.real.tx.us

Lost Maples State Natural Area
37221 RM 187, Vanderpool, TX 78885
Totality Starts at: 1:30pm CDT
Totality Lasts for: 4:25

Frio Pecan Farm Cabins & RV
144 Ranch Rd
Leakey, (830) 232-5294
www.friopecanfarm.com

Kerrville, TX
Totality Starts at: 1:32pm CDT
Totality Lasts for: 4:23
Totality Ends at: 1:36pm CDT
NWS Office: www.weather.gov/ewx
City Website: kerrvilletx.gov

Hill Country State Natural Area
10600 Bandera Creek Rd, Bandera, TX 78003
Totality Starts at: 1:31pm CDT
Totality Lasts for: 4:07

Kerrville-Schreiner Park
2385 Bandera Hwy
Kerrville, TX 78028
kerrvilletx.gov/318/Kerrville-Schreiner-Park

Riverside Nature Center
150 Francisco Lemos St
Kerrville, (830) 257-4837
riversidenaturecenter.org

Stonehenge II - Hill Country Arts Foundation
120 Point Theatre Rd S, Ingram, TX 78025
www.hcaf.com/stonehenge-ii/
Totality Lasts for: 4:25

The Museum of Western Art
1550 Bandera Hwy, Kerrville, (830) 896-2553
Art Museum
museumofwesternart.com

Tranquility Island
202 Thompson Dr
Kerrville, TX 78028
kerrvilletx.gov/1096/Tranquility-Island

Junction, TX
Totality Starts at: 1:32pm CDT
Totality Lasts for: 3:05
Totality Ends at: 1:35pm CDT
NWS Office: www.weather.gov/sjt
City Website: www.junctiontexas.com

Deer Horn Tree
1498 Main St
Junction, TX 76849
texashillcountry.com/deer-horn-tree/

Devil's Sinkhole State Natural Area
Rocksprings, TX 78880
Nature Preserve
tpwd.state.tx.us

North Llano River RV Park
2145 Main St
Junction, (325) 999-1193
www.northllanoriverrvpark.com

Schreiner Park
Junction, TX 76849
River Access
www.junctiontexas.com

South Llano River State Park
1927 Park Rd 73, Junction, TX 76849
tpwd.state.tx.us/state-parks/south-llano-river
Totality Lasts for: 3:10

Brady, TX
Totality Starts at: 1:34pm CDT
Totality Lasts for: 2:02
Totality Ends at: 1:36pm CDT
NWS Office: www.weather.gov/sjt
City Website: www.bradytx.us

D & J's Good Ole Days
109 W. Commerce St, 907 S Bridge St
Antique Store
(325) 456-9030

Earnest O. Martin Park
Memory Ln
Brady, TX 76825
www.bradytx.us/553/Parks

Heart of Texas Country Music Museum
1701 S Bridge St
Brady, (325) 597-1895
www.hillbillyhits.com

Willie Washington Park
Brady, TX 76825
City Park
www.bradytx.us/957/Willie-Washington-Park

San Antonio, TX
Totality Starts at: 1:33pm CDT
Totality Lasts for: 2:31
Totality Ends at: 1:35pm CDT
NWS Office: www.weather.gov/ewx
City Website: www.sanantonio.gov

Bamberger Nature Park
12401 Babcock Rd, San Antonio, TX 78249
www.sanantonio.gov
Totality Lasts for: 2:23

Bridgewood Park
10925 Vollmer Ln, San Antonio, TX 78254
Totality Starts at: 1:32pm CDT
Totality Lasts for: 2:31

Cathedral Rock Park
8002 Grissom Rd, San Antonio, TX 78251
www.sanantonio.gov
Totality Lasts for: 1:57

Crownridge Canyon Park
7222 Luskey Blvd, San Antonio, TX 78256
Totality Starts at: 1:33pm CDT
Totality Lasts for: 2:40

Culebra Creek Park
10919 Westwood Loop, San Antonio
Totality Starts at: 1:32pm CDT
Totality Lasts for: 2:27

Denman Estate Park
7735 Mockingbird Ln, San Antonio
Totality Starts at: 1:33pm CDT
Totality Lasts for: 1:25

Eisenhower Park
19399 NW Military Hwy, San Antonio
Totality Starts at: 1:33pm CDT
Totality Lasts for: 2:27

Friedrich Wilderness Park
21395 Milsa Dr, San Antonio
Totality Starts at: 1:33pm CDT
Totality Lasts for: 2:44

Government Canyon State Natural Area
12861 Galm Rd, San Antonio, TX 78254
tpwd.texas.gov/state-parks/government-canyon
Totality Lasts for: 2:56

Never gaze at the sun without eye protection. Only remove eye protection during 100% totality.

Helotes Ziplines
18026 Frank Madla Rd, Helotes
25-Mile Scenic Views
heloteshillcountryziplines.com

Iron Horse Canyon Recreation Center
13110 Iron Horse Way, Helotes
Totality Starts at: 1:32pm CDT
Totality Lasts for: 2:43

Natural Bridge Caverns
26495 Natural Bridge Caverns Rd, San Antonio
naturalbridgecaverns.com
Totality Lasts for: 1:08

O. P. Schnabel Park
9606 Bandera Rd, San Antonio, TX 78250
www.sanantonio.gov
Totality Lasts for: 2:19

Phil Hardberger Park
8400 NW Military Hwy, San Antonio, TX 78231
Totality Starts at: 1:33pm CDT
Totality Lasts for: 1:38

Raymond Russell Park
20644 Frontage Rd, San Antonio, TX 78257
Totality Starts at: 1:33pm CDT
Totality Lasts for: 2:39

San Antonio Aquarium
6320 Bandera Rd, Leon Valley, TX 78238
Totality Starts at: 1:33pm CDT
Totality Lasts for: 1:42

Seaworld San Antonio
10500 SeaWorld Dr, San Antonio, TX 78251
Totality Starts at: 1:33pm CDT
Totality Lasts for: 2:10

Senator Frank L. Madla Natural Area
9780 Menchaca Rd, Helotes, TX 78023
madlapark.org
Totality Lasts for: 2:48

Six Flags Fiesta Texas
San Antonio, TX 78257
sixflags.com/fiestatexas
Totality Lasts for: 2:30

Tom Slick Park
7400 Texas 151 Access Rd, San Antonio
Totality Starts at: 1:33pm CDT
Totality Lasts for: 1:18

Urban Air Trampoline and Adventure Park
11791 Bandera Rd Suite A, San Antonio
Totality Starts at: 1:33pm CDT
Totality Lasts for: 2:30

Boerne, TX
Totality Starts at: 1:32pm CDT
Totality Lasts for: 3:34
Totality Ends at: 1:36pm CDT
NWS Office: www.weather.gov/ewx
City Website: www.ci.boerne.tx.us

Boerne City Lake Park
1 City Lk Rd, Boerne, TX 78006
Totality Starts at: 1:32pm CDT
Totality Lasts for: 3:47

Boerne City Park
106 City Park Rd, Boerne, TX 78006
Totality Starts at: 1:32pm CDT
Totality Lasts for: 3:29

Cascade Caverns
226 Cascade Cavern
Boerne, TX 78015
www.cascadecaverns.com

Cave Without A Name
325 Kreutzberg Rd
Boerne, TX 78006
www.cavewithoutaname.com

Guadalupe River State Park
3350 Park Rd 31, Texas 78070
tpwd.texas.gov/state-parks/guadalupe-river
Totality Lasts for: 3:08

Kinderpark
Boerne, TX 78006
Totality Starts at: 1:32pm CDT
Totality Lasts for: 3:32

Northrup Park
37550 Frontage Rd, Boerne, TX 78006
Totality Starts at: 1:32pm CDT
Totality Lasts for: 3:39

Llano, TX
Totality Starts at: 1:34pm CDT
Totality Lasts for: 4:23
Totality Ends at: 1:38pm CDT
NWS Office: www.weather.gov/ewx
City Website: www.cityofllano.com

Badu Park
Llano, TX 78643
Totality Starts at: 1:34pm CDT
Totality Lasts for: 4:22

Enchanted Rock State Natural Area
16710 Ranch Rd 965, Fredericksburg
Totality Starts at: 1:33pm CDT
Totality Lasts for: 4:24

Llano City Rv Park
100 Robinson Park Dr
Llano, TX 78643
www.llanocityrvpark.com

Llano County Museum
310 Bessemer Ave
Llano, TX 78643
llanomuseum.org

Longhorn Cavern State Park
6211 Park Road 4 S, Burnet, TX 78611
www.visitlonghorncavern.com
Totality Lasts for: 4:21

Inks Lake State Park
3630 Park Rd 4 W, Burnet, TX 78611
tpwd.texas.gov/state-parks/inks-lake
Totality Lasts for: 4:22

Austin, TX
Totality Starts at: 1:36pm CDT
Totality Lasts for: 2:18
Totality Ends at: 1:38pm CDT
NWS Office: www.weather.gov/ewx
City Website: www.austintexas.gov

Austin Nature & Science Center
2389 Stratford Dr
Austin, TX 78746
www.austintexas.gov

Barton Creek Habitat Preserve
Austin, TX 78733
Totality Starts at: 1:35pm CDT
Totality Lasts for: 2:43

Brushy Creek Lake Park
3300 Brushy Creek Rd, Cedar Park, TX 78613
Totality Starts at: 1:35pm CDT
Totality Lasts for: 3:06

Bull Creek District Park
6701 Lakewood Dr, Austin, TX 78731
Totality Starts at: 1:35pm CDT
Totality Lasts for: 2:38

Bullock Texas State History Museum
1800 Congress Ave.
Austin, TX 78701
www.thestoryoftexas.com

Cathedral of Junk
4422 Lareina Dr, Austin, TX 78745
Totality Starts at: 1:36pm CDT
Totality Lasts for: 1:26

Commons Ford Ranch Metropolitan Park
614 N Commons Ford Rd, Austin, TX 78733
Totality Starts at: 1:35pm CDT
Totality Lasts for: 2:54

Doug Sahm Hill Summit
Doug Sahm Hill Path, Butler Metro Park
Austin, TX 78704
austinparks.org/butler-park

Emma Long Metropolitan Park
1600 City Park Rd, Austin, TX 78730
Totality Starts at: 1:35pm CDT
Totality Lasts for: 2:41

Inner Space Cavern
4200 S I-35 Frontage Rd, Georgetown
Totality Starts at: 1:36pm CDT
Totality Lasts for: 3:12

Lou Neff Point
Ann and Roy Butler Trail, Austin
Totality Starts at: 1:36pm CDT
Totality Lasts for: 1:56

Mansfield Dam Park
4370 Mansfield Dam Park Rd., Austin
Totality Starts at: 1:35pm CDT
Totality Lasts for: 3:08

Never gaze at the sun without eye protection. Only remove eye protection during 100% totality.

Mayfield Park and Nature Preserve
3505 W 35th St, Austin, TX 78703
Totality Starts at: 1:35pm CDT
Totality Lasts for: 2:17

Mueller Lake Park
4550 Mueller Blvd, Austin
Totality Starts at: 1:36pm CDT
Totality Lasts for: 1:41

Museum of the Weird
412 E 6th St
Austin, TX 78701
www.museumoftheweird.com

Old Settlers Park
3300 E Palm Valley Blvd, Round Rock
Totality Starts at: 1:36pm CDT
Totality Lasts for: 2:46

Pease District Park
1100 Kingsbury St, Austin, TX 78703
Totality Starts at: 1:36pm CDT
Totality Lasts for: 1:59

Pedernales Falls State Park
2585 Park Rd 6026, Johnson City, TX 78636
tpwd.texas.gov/state-parks/pedernales-falls
Totality Lasts for: 3:43

River Place Nature Trail Canyon Trailhead
4998 River Pl Blvd, Austin, TX 78730
Totality Starts at: 1:35pm CDT
Totality Lasts for: 2:54

Sculpture Falls
Barton Creek Greenbelt Trail, Austin
Totality Starts at: 1:35pm CDT
Totality Lasts for: 2:17

St. Edward's Park
7301 Spicewood Springs Rd, Austin, TX 78759
Totality Starts at: 1:35pm CDT
Totality Lasts for: 2:48

Texas Capitol
1100 Congress Ave., Austin, TX 78701
Totality Starts at: 1:36pm CDT
Totality Lasts for: 1:49

Walnut Creek Metropolitan Park
12138 N Lamar Blvd, Austin, TX 78753
Totality Starts at: 1:36pm CDT
Totality Lasts for: 2:23

Zilker Metropolitan Park
Austin, TX 78746
Totality Starts at: 1:36pm CDT
Totality Lasts for: 1:58

Taylor, TX
Totality Starts at: 1:37pm CDT
Totality Lasts for: 1:51
Totality Ends at: 1:38pm CDT
NWS Office: www.weather.gov/ewx
City Website: www.ci.taylor.tx.us

Blackland Farms RV Park
1800 Co Rd 374, Taylor, TX 76574
Totality Starts at: 1:36pm CDT
Totality Lasts for: 2:27

Bull Branch Park
904 Dellinger Dr, Taylor, TX 76574
Totality Starts at: 1:37pm CDT
Totality Lasts for: 2:06

Fannie Robinson Park
206 S Dolan St, Taylor, TX 76574
Totality Starts at: 1:37pm CDT
Totality Lasts for: 1:43

Four Winds RV Park
408 Carlos G Parker Blvd SE, Taylor
Totality Starts at: 1:37pm CDT
Totality Lasts for: 1:32

Murphy Park
1600 Veterans Dr, Taylor, TX 76574
Totality Starts at: 1:37pm CDT
Totality Lasts for: 2:00

Georgetown, TX
Totality Starts at: 1:36pm CDT
Totality Lasts for: 3:16
Totality Ends at: 1:39pm CDT
NWS Office: www.weather.gov/ewx
City Website: georgetown.org

Berry Springs Park and Preserve
1801 Co Rd 152, Georgetown, TX 78626
Totality Starts at: 1:36pm CDT
Totality Lasts for: 3:19

Central Texas RV Parks
349 FM1105
Georgetown, TX 78626
centraltexasrvparks.com

Fritz Park
400 Park Ave, Hutto, TX 78634
Totality Starts at: 1:36pm CDT
Totality Lasts for: 2:28

Garey Park
6450 RTM Rd 2243, Georgetown, TX 78628
Totality Starts at: 1:35pm CDT
Totality Lasts for: 3:26

Overlook Park
508-598 Lake Overlook Rd, Georgetown
Totality Starts at: 1:36pm CDT
Totality Lasts for: 3:29

Jim Hogg Park
500 Jim Hogg Rd
Georgetown, TX 78633
recreation.gov

Rivery Park
1125 Woodlawn Ave, Georgetown, TX 78626
Totality Starts at: 1:36pm CDT
Totality Lasts for: 3:19

Lampasas, TX
Totality Starts at: 1:35pm CDT
Totality Lasts for: 4:24
Totality Ends at: 1:39pm CDT
NWS Office: www.weather.gov/fwd
City Website: www.lampasas.org

Colorado Bend State Park
2236 Park Hill Dr, Bend, TX 76824
tpwd.texas.gov/state-parks/colorado-bend
Totality Lasts for: 4:19

Hanna Springs Sculpture Garden
501 E North Ave
Lampasas, TX 76550
www.lafta.org

Hancock Springs Park
Lampasas, TX 76550
Totality Starts at: 1:35pm CDT
Totality Lasts for: 4:24

Lampasas County Museum
303 S Western Ave
Lampasas, TX 76550
lampasasmuseum.org

Killeen, TX
Totality Starts at: 1:36pm CDT
Totality Lasts for: 4:17
Totality Ends at: 1:40pm CDT
NWS Office: www.weather.gov/fwd
City Website: www.killeentexas.gov

Carl Levin Park
400 Millers Crossing, Harker Heights
Totality Starts at: 1:36pm CDT
Totality Lasts for: 4:11

Community Center Park
2201 E Veterans Memorial Blvd
Killeen, TX 76543
www.killeentexas.gov

Copperas Cove City Park
1206 W Avenue B, Copperas Cove, TX 76522
Totality Starts at: 1:36pm CDT
Totality Lasts for: 4:22

Conder Park
810 Conder St
Killeen, TX 76541
www.killeentexas.gov

Dana Peak Park
3800 Comanche Gap Rd, Harker Heights
Totality Starts at: 1:36pm CDT
Totality Lasts for: 4:05

Harker Heights Community Park
1605 Knight's Way, Harker Heights, TX 76548
Totality Starts at: 1:36pm CDT
Totality Lasts for: 4:09

Lions Club Park
1700 E Stan Schlueter Loop
Killeen, TX 76542
www.killeentexas.gov

Long Branch Park
1101 Branch Dr
Killeen, TX 76543
www.killeentexas.gov

Never gaze at the sun without eye protection. Only remove eye protection during 100% totality.

Purser Family Park
100 Mountain Lion Rd, Harker Heights
Totality Starts at: 1:36pm CDT
Totality Lasts for: 4:11

Stillhouse Park
4050 Simmons Rd, Belton, TX 76513
Totality Starts at: 1:36pm CDT
Totality Lasts for: 4:00

Temple, TX
Totality Starts at: 1:37pm CDT
Totality Lasts for: 3:49
Totality Ends at: 1:40pm CDT
NWS Office: www.weather.gov/fwd
City Website: www.templetx.gov

Bell County Museum
201 N Main St, Belton, TX 76513
Totality Starts at: 1:36pm CDT
Totality Lasts for: 3:55

Ferguson Park
1203 E Adams Ave, Temple, TX 76501
Totality Starts at: 1:37pm CDT
Totality Lasts for: 3:43

Heritage Park
1502 Park Ave, Belton, TX 76513
Totality Starts at: 1:36pm CDT
Totality Lasts for: 3:54

James Wilson Park
1909 Curtis B Elliot Dr, Temple
Totality Starts at: 1:37pm CDT
Totality Lasts for: 3:41

Miller Springs Nature Center
1473 Farm-To-Market Rd 2271, Belton
Totality Starts at: 1:36pm CDT
Totality Lasts for: 4:02

Rogers Park
Belton, TX 76513
Totality Starts at: 1:36pm CDT
Totality Lasts for: 4:06

Temple Lake Park
14190 FM2305, Belton, TX 76513
Totality Starts at: 1:36pm CDT
Totality Lasts for: 4:05

Temple Railroad & Heritage Museum
315 W Avenue B
Temple, TX 76501
www.templerrhm.org

West Temple Community Park
8420 W Adams Ave, Temple, TX 76502
Totality Starts at: 1:37pm CDT
Totality Lasts for: 3:58

Whistle Stop Park
58 S 11th St
Temple, TX 76501
www.templeparks.com

Winkler Park
11740 Winkler Park Rd, Moody, TX 76557
Totality Starts at: 1:37pm CDT
Totality Lasts for: 4:12

Yettie Polk Park
101 S Davis St, Belton, TX 76513
Totality Starts at: 1:36pm CDT
Totality Lasts for: 3:55

Gatesville, TX
Totality Starts at: 1:36pm CDT
Totality Lasts for: 4:24
Totality Ends at: 1:41pm CDT
NWS Office: www.weather.gov/fwd
City Website: www.gatesvilletx.com

Coryell Museum Historical Center
718 E Main St, Gatesville, TX 76528
Totality Starts at: 1:36pm CDT
Totality Lasts for: 4:23

Faunt Le Roy RV Park
S 7th St
Gatesville, TX 76528
visitgatesvilletx.com/faunt-le-roy-rv-parking

Waco, TX
Totality Starts at: 1:37pm CDT
Totality Lasts for: 4:15
Totality Ends at: 1:42pm CDT
NWS Office: www.weather.gov/fwd
City Website: www.waco-texas.com

Airport Beach Park
4600 Skeet Eason Rd, Waco, TX 76708
Totality Starts at: 1:37pm CDT
Totality Lasts for: 4:18

Cameron Park
2601 N University Parks Dr
Waco, TX 76708
www.waco-texas.com

Cameron Park Zoo
1701 N 4th St
Waco, TX 76707
www.cameronparkzoo.com

Carleen Bright Arboretum
9001 Bosque Blvd b, Woodway, TX 76712
Totality Starts at: 1:37pm CDT
Totality Lasts for: 4:14

Dr Pepper Museum
300 S 5th St
Waco, TX 76701
www.drpeppermuseum.com

Fort Parker State Park
194 Park Rd, Mexia, TX 76667
tpwd.texas.gov/state-parks/fort-parker
Totality Lasts for: 3:08

Lake Waco Wetlands
1752 Eichelberger Crossing, China Spring
Totality Starts at: 1:37pm CDT
Totality Lasts for: 4:19

Mayborn Museum Complex
1300 S University Parks Dr
Waco, TX 76706
www.baylor.edu/mayborn/

Mother Neff State Park
1921 State Park Rd 14, Moody, TX 76557
tpwd.texas.gov/state-parks/mother-neff
Totality Lasts for: 4:16

Pecan Bottom Park
Cameron Park Dr
Waco, TX 76707
www.waco-texas.com

Texas Ranger Hall of Fame & Museum
100 Texas Ranger Trail
Waco, TX 76706
www.texasranger.org

Tonkawa Falls City Park
524-550 E 4th St, Crawford, TX 76638
Totality Starts at: 1:37pm CDT
Totality Lasts for: 4:21

Waco Mammoth National Monument
6220 Steinbeck Bend Dr, Waco, TX 76708
Totality Starts at: 1:38pm CDT
Totality Lasts for: 4:16

Waco Suspension Bridge
101 N University Parks Dr
Waco, TX 76701
www.waco-texas.com

Woodway Park
Estates Dr, Waco, TX 76712
Totality Starts at: 1:37pm CDT
Totality Lasts for: 4:15

Hillsboro, TX
Totality Starts at: 1:38pm CDT
Totality Lasts for: 4:22
Totality Ends at: 1:43pm CDT
NWS Office: www.weather.gov/fwd
City Website: www.hillsborotx.org

Cedar Creek Park
Whitney, TX 76692
Totality Starts at: 1:38pm CDT
Totality Lasts for: 4:15

Cedron Creek Park
557 FM 1713, Whitney, TX 76692
Totality Starts at: 1:38pm CDT
Totality Lasts for: 4:15

Hillsboro City Park
N Pleasant St
Hillsboro, TX 76645
www.hillsborotx.org/city-park

Never gaze at the sun without eye protection. Only remove eye protection during 100% totality.

History Of West Museum
112 E Oak St, West, TX 76691
Totality Starts at: 1:38pm CDT
Totality Lasts for: 4:20

Lake Whitney State Park
433 FM1244, Whitney, TX 76692
tpwd.texas.gov/state-parks/lake-whitney
Totality Lasts for: 4:20

McCown Valley Park
283 McCown Valley Park Road, Whitney
Totality Starts at: 1:38pm CDT
Totality Lasts for: 4:17

Meridian State Park
173 Park Road #7, Meridian, TX 76665
tpwd.texas.gov/state-parks/meridian
Totality Lasts for: 4:01

Plowman Creek Park
15188 FM 56, Kopperl, TX 76652
Totality Starts at: 1:38pm CDT
Totality Lasts for: 4:02

Roadside America Museum
212 E Elm St
Hillsboro, TX 76645
www.roadsideamericatx.com

Texas Heritage Museum
112 Lamar Dr
Hillsboro, TX 76645
www.hillcollege.edu/museum/

Texas Through Time
110 N Waco St
Hillsboro, TX 76645
texasthroughtime.org

Wallace Park
TX-81
Hillsboro, TX 76645
wallace-park.edan.io/

Cleburne, TX
Totality Starts at: 1:39pm CDT
Totality Lasts for: 3:42
Totality Ends at: 1:42pm CDT
NWS Office: www.weather.gov/fwd
City Website: www.cleburne.net

Carver Park
600 Park St, Cleburne, TX 76031
Totality Starts at: 1:39pm CDT
Totality Lasts for: 3:42

Chisholm Trail Park
TX-174, Blum, TX 76627
Totality Starts at: 1:38pm CDT
Totality Lasts for: 3:56

Cleburne State Park
5800 Park Rd 21, Cleburne, TX 76033
tpwd.texas.gov/state-parks/cleburne
Totality Lasts for: 3:34

Dinosaur Valley State Park
Glen Rose, TX 76043
tpwd.texas.gov/state-parks/dinosaur-valley
Totality Lasts for: 3:00

Fossil Rim Wildlife Center
2299 Co Rd 2008, Glen Rose, TX 76043
Totality Starts at: 1:38pm CDT
Totality Lasts for: 3:15

Gone With the Wind Museum
305 E 2nd St
Cleburne, TX 76031
www.gwtwremembered.com

Granbury City Beach Park
505 E Pearl St, Granbury, TX 76048
Totality Starts at: 1:39pm CDT
Totality Lasts for: 2:14

Ham Creek Park
6957 FM 916, Rio Vista, TX 76093
Totality Starts at: 1:38pm CDT
Totality Lasts for: 3:53

Hulen Park
301 W Westhill Dr
Cleburne, TX 76033
www.cleburne.net

Kimball Bend Park
3351TX, 174, Kopperl, TX 76652
Totality Starts at: 1:38pm CDT
Totality Lasts for: 3:56

Layland Museum
201 Caddo St
Cleburne, TX 76031
laylandmuseum.net

Texas Children's Museum
201 TX-174, Rio Vista, TX 76093
Totality Starts at: 1:38pm CDT
Totality Lasts for: 3:55

The Chisholm Trail Outdoor Museum;
Big Bear Native American Museum
101 Chisholm Trail, Cleburne, TX 76033
www.jcchisholmtrail.com

Corsicana, TX
Totality Starts at: 1:40pm CDT
Totality Lasts for: 4:10
Totality Ends at: 1:44pm CDT
NWS Office: www.weather.gov/fwd
www.cityofcorsicana.com

Fairfield Lake State Park
123 State Park Rd. 64, Fairfield, TX 75840
tpwd.texas.gov/state-parks/fairfield-lake
Totality Lasts for: 2:20

Fullerton-Garitty Park
3201 McKnight Ln, Corsicana, TX 75110
Totality Starts at: 1:40pm CDT
Totality Lasts for: 4:13

Jester Park
Corsicana, TX 75110
Totality Starts at: 1:40pm CDT
Totality Lasts for: 4:10

Pearce Collections Museum
3100 W Collin St
Corsicana, TX 75110
www.pearcemuseum.com

Petroleum Park
416 S 12th St
Corsicana, TX 75110
www.cityofcorsicana.com

Pioneer Village
912 W Park Ave
Corsicana, TX 75110
www.cityofcorsicana.com/995/Pioneer-Village

Ennis, TX
Totality Starts at: 1:39pm CDT
Totality Lasts for: 4:22
Totality Ends at: 1:44pm CDT
NWS Office: www.weather.gov/fwd
City Website: www.ennistx.gov

Bluebonnet Park
201 US-287
Ennis, TX 75119
www.ennistx.gov

Ennis Railroad & Cultural Heritage Museum
105 NE Main St
Ennis, TX 75119
www.ennistx.gov

High View Park
262 High View Marina Rd, Ennis, TX 75119
Totality Starts at: 1:39pm CDT
Totality Lasts for: 4:22

Kachina Prairie Park
1816 W Baldridge St
Ennis, TX 75119
www.ennistx.gov

Lake Clark Park
2100 W Baldridge St
Ennis, TX 75119
www.ennistx.gov

Lions Park
1200 W Lampasas St
Ennis, TX 75119
www.ennistx.gov

Waxahachie Creek Park
930 Bozek Ln, Ennis, TX 75119
Totality Starts at: 1:39pm CDT
Totality Lasts for: 4:22

Never gaze at the sun without eye protection. Only remove eye protection during 100% totality.

Waxahachie, TX
Totality Starts at: 1:39pm CDT
Totality Lasts for: 4:17
Totality Ends at: 1:44pm CDT
NWS Office: www.weather.gov/fwd
City Website: www.waxahachie.com

Boat Dock Park
111 Lakeshore Dr, Waxahachie, TX 75165
Totality Starts at: 1:39pm CDT
Totality Lasts for: 4:20

Brown Singleton Park
847 Farley St
Waxahachie, TX 75165
www.waxahachie.com

Chapman Park
1805 Alexander Dr
Waxahachie, TX 75165
www.waxahachie.com

Ellis County Museum
201 S College St
Waxahachie, TX 75165
www.elliscountymuseum.org

Getzendaner Park
400 S Grand Ave
Waxahachie, TX 75165
www.waxahachie.com

Lee Penn Park
404 Getzendaner Street
Waxahachie, TX 75165
www.waxahachie.com

Mockingbird Nature Park
1361 Onward Rd, Midlothian, TX 76065
Totality Starts at: 1:39pm CDT
Totality Lasts for: 4:02

Munster Mansion
3636 FM813
Waxahachie, TX 75165
www.munstermansion.com

Poston Gardens
2000 Civic Center Ln
Waxahachie, TX 75165
tulipalooza.org

Athens, TX
Totality Starts at: 1:41pm CDT
Totality Lasts for: 3:17
Totality Ends at: 1:44pm CDT
NWS Office: www.weather.gov/fwd
City Website: www.athenstx.gov

211 Gallery - Art Matters
211 N Palestine St
Athens, TX 75751
artgallery211.net

Coleman Park
1950 E College St
Athens, TX 75751
www.athenstx.gov

East Texas Arboretum
1601 Patterson Rd
Athens, TX 75751
www.easttexasarboretum.org

Gun Barrel City Park
301 Municipal Dr, Gun Barrel City, TX 75156
Totality Starts at: 1:41pm CDT
Totality Lasts for: 4:06

Kiwanis Park
406 S Prairieville St
Athens, TX 75751
www.athenstx.gov

Log Cabin City Park
Alamo Rd, Log Cabin, TX 75148
Totality Starts at: 1:41pm CDT
Totality Lasts for: 3:45

New York, Texas ZipLine Adventures
7290 Co Rd 4328, Larue, TX 75770
Totality Starts at: 1:42pm CDT
Totality Lasts for: 2:24

Purtis Creek State Park
14225 FM 316 N, Eustace, TX 75124
tpwd.texas.gov/state-parks/purtis-creek
Totality Lasts for: 4:00

Rocky Ridge Drive-Thru Safari
3350 FM1256, Eustace, TX 75124
Totality Starts at: 1:41pm CDT
Totality Lasts for: 3:56

Texas Freshwater Fisheries Center
5550 FM2495, Athens, TX 75752
Totality Starts at: 1:41pm CDT
Totality Lasts for: 3:07

Tom Finley Park
TX-334, Gun Barrel City, TX 75156
Totality Starts at: 1:40pm CDT
Totality Lasts for: 4:09

Tyler, TX
Totality Starts at: 1:43pm CDT
Totality Lasts for: 2:13
Totality Ends at: 1:45pm CDT
NWS Office: www.weather.gov/shv
City Website: www.cityoftyler.org

Bergfeld Park
1510 S College Ave
Tyler, TX 75702
www.cityoftyler.org

Caldwell Zoo
2203 W MLK Jr Blvd
Tyler, TX 75702
www.caldwellzoo.org

Cotton Belt Depot Museum
210 E Oakwood St
Tyler, TX 75702
www.cottonbeltdepotmuseum.com

Darden Harvest Park
202 Cannery Row, Lindale, TX 75771
Totality Starts at: 1:42pm CDT
Totality Lasts for: 3:08

Goodman-LeGrand House & Museum
624 N Broadway Ave
Tyler, TX 75702
www.cityoftyler.org

Northside Park
2301 W NW Loop 323, Tyler, TX 75702
Totality Starts at: 1:43pm CDT
Totality Lasts for: 2:19

Old Mill Museum
2900 S Main St, Lindale, TX 75771
Totality Starts at: 1:43pm CDT
Totality Lasts for: 2:57

Southside Park
455 Shiloh Rd, Tyler, TX 75703
Totality Starts at: 1:43pm CDT
Totality Lasts for: 1:07

The Discovery Science Place
308 N Broadway Ave
Tyler, TX 75702
www.discoveryscienceplace.org

The Earth And Space Science Center
1411 E Lake St
Tyler, TX 75701
sciencecenter.tjc.edu

Tyler Museum of Art
1300 S Mahon Ave
Tyler, TX 75701
www.tylermuseum.org

Tyler State Park
789 Park Rd 16, Tyler, TX 75706
tpwd.texas.gov/state-parks/tyler
Totality Lasts for: 2:34

Dallas-Fort Worth
Totality Starts at: 1:40pm CDT
Totality Lasts for: D 3:50, FW 2:35
Totality Ends at: 1:44pm CDT
NWS Office: www.weather.gov/fwd
dallascityhall.com, fortworthtexas.gov

African American Museum of Dallas
3536 Grand Ave
Dallas, TX 75210
www.aamdallas.org

Amon Carter Museum of American Art
3501 Camp Bowie Blvd, Fort Worth, TX 76107
Totality Starts at: 1:40pm CDT
Totality Lasts for: 2:28

Cedar Hill State Park
1570 FM1382, Cedar Hill, TX 75104
tpwd.texas.gov/state-parks/cedar-hill
Totality Lasts for: 3:49

Cobb Park
2700 Cobb Park Dr W, Fort Worth, TX 76105
Totality Starts at: 1:40pm CDT
Totality Lasts for: 2:54

Never gaze at the sun without eye protection. Only remove eye protection during 100% totality.

Dallas Arboretum and Botanical Garden
8525 Garland Rd, www.dallasarboretum.org
Totality Starts at: 1:40pm CDT
Totality Lasts for: 3:51

Dallas Museum of Art
1717 N Harwood St
Dallas, TX 75201
www.dma.org

Dallas Zoo
650 S R.L. Thornton Fwy
Dallas, TX 75203
www.dallaszoo.com

Dealey Plaza
www.jfk.org, Dallas, TX 75202
Totality Starts at: 1:40pm CDT
Totality Lasts for: 3:47

Fort Worth Botanic Garden
3220 Botanic Garden Blvd, Fort Worth
Totality Starts at: 1:40pm CDT
Totality Lasts for: 2:31

Fort Worth's Flatiron Building
1000 Houston St
Fort Worth, TX 76102
fortworthflatironbuilding.com

Fort Worth Water Gardens
1502 Commerce St, Fort Worth, TX 76102
Totality Starts at: 1:40pm CDT
Totality Lasts for: 2:38

Frontiers of Flight Museum, Dallas Love Field
6911 Lemmon Ave, www.flightmuseum.com
Totality Starts at: 1:40pm CDT
Totality Lasts for: 3:36

Giant Eyeball
1601 Main St
Dallas, TX 75201
artandseek.org

John F. Kennedy Memorial Plaza
646 Main St
Dallas, TX 75202
www.jfk.org

John Wayne: An American Experience
2501 Rodeo Plaza, Fort Worth, johnwayne.com
Totality Starts at: 1:40pm CDT
Totality Lasts for: 2:19

Kidd Springs Park
1003 Cedar Hill Ave
Dallas, TX 75208
dallasparks.org

Kimbell Art Museum
3333 Camp Bowie Blvd, Fort Worth, TX 76107
Totality Starts at: 1:40pm CDT
Totality Lasts for: 2:28

Klyde Warren Park
2012 Woodall Rodgers Fwy
Dallas, TX 75201
klydewarrenpark.org

Lake Cliff Park
300 E Colorado Blvd
Dallas, TX 75201
www.dallasparks.org

Main Street Garden Park
1902 Main St
Dallas, TX 75201
mainstreetgarden.org

Museum of Illusions Dallas
701 Ross Ave
Dallas, TX 75202
moidallas.com

Old City Park
1515 S Harwood St
Dallas, TX 75215
www.oldcityparkdallas.org

Perot Museum of Nature and Science
2201 N Field St, Dallas, TX 75201
Totality Starts at: 1:40pm CDT
Totality Lasts for: 3:47

Pioneer Plaza
1428 Young St
Dallas, TX 75202
www.dallasparks.org

Ripley's Believe It or Not!
601 E Palace Pkwy, Grand Prairie, TX 75050
Totality Starts at: 1:40pm CDT
Totality Lasts for: 3:28

Reunion Tower
300 Reunion Blvd E
Dallas, TX 75207
www.reuniontower.com

Reverchon Park
3505 Maple Ave
Dallas, TX 75219
www.dallasparks.org

Santa Fe Trestle Trail Bridge
Trestle Trail, Dallas, TX 75203
Totality Starts at: 1:40pm CDT
Totality Lasts for: 3:52

Six Flags Over Texas
Arlington, TX, www.sixflags.com
Totality Starts at: 1:40pm CDT
Totality Lasts for: 3:20

Texas Discovery Gardens
3601 Martin Luther King Jr Blvd
Dallas, TX 75210
www.txdg.org

The Dallas World Aquarium
1801 N Griffin St
Dallas, TX 75202
www.dwazoo.com

Trinity Overlook Park
110 W Commerce St
Dallas, TX 75205
www.trinityrivercorridor.com

The Sixth Floor Museum at Dealey Plaza
411 Elm St, Dallas, www.jfk.org
Totality Starts at: 1:40pm CDT
Totality Lasts for: 3:47

White Rock Lake Park
8300 E Lawther Dr
Dallas, TX 75218
www.whiterocklake.org

Sulphur Springs, TX
Totality Starts at: 1:42pm CDT
Totality Lasts for: 4:20
Totality Ends at: 1:47pm CDT
NWS Office: www.weather.gov/fwd
www.sulphurspringstx.org

Buford Park
733 Connally St
Sulphur Springs, TX 75482
www.sulphurspringstx.org

Coleman Park
Sulphur Springs, TX 75482
Totality Starts at: 1:42pm CDT
Totality Lasts for: 4:20

Cooper Lake State Park South Sulphur Unit
1690 FM 3505, Sulphur Springs, TX 75482
tpwd.texas.gov/state-parks/cooper-lake
Totality Lasts for: 4:19

Hopkins County Historical Society & Museum
416 Jackson St N
Sulphur Springs, TX 75482
www.hopkinscountymuseum.org

Hopkins County Veteran's Memorial
116 Oak Ave
Sulphur Springs, TX 75482
www.sulphurspringstx.org

Pacific Park
413 Beckham St W
Sulphur Springs, TX 75482
www.sulphurspringstx.org

Southwest Dairy Museum
1210 Houston St
Sulphur Springs, TX 75482
www.southwestdairyfarmers.com

Paris, TX
Totality Starts at: 1:44pm CDT
Totality Lasts for: 4:00
Totality Ends at: 1:48pm CDT
NWS Office: www.weather.gov/fwd
City Website: www.paristexas.gov

Bonham State Park
1363 State Park 24, Bonham, TX 75418
tpwd.texas.gov/state-parks/bonham
Totality Lasts for: 3:14

Bywaters Park
300 S Main St
Paris, TX 75460
www.paristexas.gov

Caddo National Grasslands
Honey Grove, TX 75446
Totality Starts at: 1:43pm CDT
Totality Lasts for: 3:05

Never gaze at the sun without eye protection. Only remove eye protection during 100% totality.

Eiffel Tower Paris Texas
2025 S Collegiate Dr, Paris, TX 75460
65ft Tower, Tourist Attraction
www.paristexas.com

Paris Drag Strip
4417 US-82
Paris, TX 75462
www.theparisdragstrip.com

Pat Mayse Lake Recreation Area
Powderly, TX 75473
Totality Starts at: 1:44pm CDT
Totality Lasts for: 3:41

Pat Mayse West Campground
CR 35800
Powderly, TX 75473
recreation.gov

Sam Bell Maxey House State Historic Site
812 S Church St
Paris, TX 75460
www.visitsambellmaxeyhouse.com

Wade Park
2400 E Price St
Paris, TX 75460
www.paristexas.gov

Mt Pleasant, TX
Totality Starts at: 1:44pm CDT
Totality Lasts for: 3:56
Totality Ends at: 1:48pm CDT
NWS Office: www.weather.gov/shv
City Website: mpcity.net

Daingerfield State Park
455 Park Rd 17, Daingerfield, TX 75638
tpwd.texas.gov/state-parks/daingerfield
Totality Lasts for: 2:52

Dellwood Park
726 E Ferguson Rd
Mt Pleasant, TX 75455
www.mpcity.net

Lake Bob Sandlin State Park
341 State, Park Rd 2117, Pittsburg, TX
tpwd.texas.gov/state-parks/lake-bob-sandlin
Totality Lasts for: 3:55

Mid America Flight Museum
602 Mike Hall Pkwy
Mt Pleasant, TX 75455
www.midamericaflightmuseum.com

Tankersley Lake Park
Mt Pleasant, TX
Totality Starts at: 1:44pm CDT
Totality Lasts for: 3:56

Walleye Park
County Rd SE 3122, Mt Vernon, TX 75457
Totality Starts at: 1:43pm CDT
Totality Lasts for: 4:01

Texarkana, TX / AR
Totality Starts at: 1:46pm CDT
Totality Lasts for: 2:35
Totality Ends at: 1:49pm CDT
NWS Office: www.weather.gov/shv
ci.texarkana.tx.us

Ace of Clubs House
420 Pine St
Texarkana, TX 75501
www.texarkanamuseums.org

Atlanta State Park
927 Park Rd 42, Atlanta, TX 75551
tpwd.texas.gov/state-parks/atlanta
Totality Lasts for: 2:20

Bringle Lake Park East
3650 Shilling Rd, Texarkana, TX 75503
Totality Starts at: 1:46pm CDT
Totality Lasts for: 2:55

Bringle Lake Park West
7602 University Ave, Texarkana, TX 75503
Totality Starts at: 1:46pm CDT
Totality Lasts for: 2:56

Clear Springs Park
64 Clear Springs Rd, Texarkana, TX 75501
Totality Starts at: 1:46pm CDT
Totality Lasts for: 2:40

Ferguson Park
3000 Texas Blvd
Texarkana, TX 75503
www.ci.texarkana.tx.us

Four States Auto Museum
217 Laurel St
Texarkana, AR 71854
www.fourstatesautomuseum.org

Herron Creek Park
Maud, TX 75567
Totality Starts at: 1:46pm CDT
Totality Lasts for: 2:50

Rocky Point Park
Rocky Point Park Rd, Queen City, TX 75572
Totality Starts at: 1:46pm CDT
Totality Lasts for: 2:12

Spring Lake Park
4303 N Park Rd
Texarkana, TX 75501
www.ci.texarkana.tx.us

Oklahoma

Idabel, OK
Totality Starts at: 1:45pm CDT
Totality Lasts for: 4:18
Totality Ends at: 1:49pm CDT
NWS Office: www.weather.gov/shv
City Website: www.idabel-ok.gov

Hugo Lake Park
Highway 70, Sawyer, OK 74756
Totality Starts at: 1:45pm CDT
Totality Lasts for: 3:30

Museum of the Red River
812 E Lincoln Rd, Idabel, OK 74745
Museum
www.museumoftheredriver.org

Raymond Gary State Park
1119 OK-209, Fort Towson, OK 74735
travelok.com/state-parks/raymond-gary-state-park
Totality Lasts for: 3:48

Broken Bow, OK
Totality Starts at: 1:45pm CDT
Totality Lasts for: 4:16
Totality Ends at: 1:50pm CDT
NWS Office: www.weather.gov/shv
City Website: cityofbrokenbow.com

Beaver's Bend Mining Company
9221 N US Hwy 259, Hochatown, OK 74728
Amusement Park
beaversbendminingcompany.com

Beavers Bend State Park and Nature Center
4350 OK-259A, Broken Bow, OK 74728
www.travelok.com
Totality Lasts for: 4:13

Beavers Bend Wildlife Museum
6594 N US Hwy 259, Broken Bow, OK 74728
Totality Starts at: 1:45pm CDT
Totality Lasts for: 4:13

Bigfoot Speedway
52 Stevens Gap Rd
Hochatown, OK 74728
www.bigfootspeedway.com

Broken Bow Overlook At Beavers Bend SP
Broken Bow, OK 74728
Totality Starts at: 1:46pm CDT
Totality Lasts for: 4:11

Lost Rapids Park
RT 1 BOX 400, 9615, Valliant, OK 74764
recreation.gov
Totality Lasts for: 3:40

Rugaru Adventures Ziplining Tour
2658 Stevens Gap Rd
Broken Bow, OK 74728
rugaruadventures.com

Never gaze at the sun without eye protection. Only remove eye protection during 100% totality.

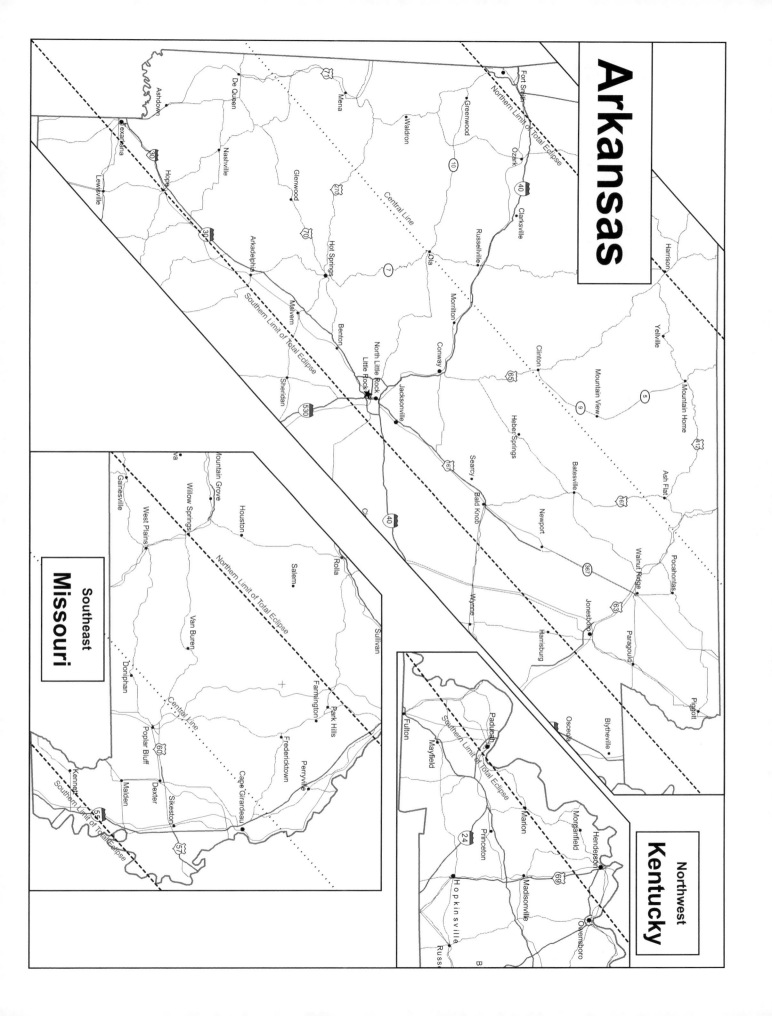

Featured Places to Eat

Bo's Breakfast and Bar-B-Q in Sikeston, welcomes diners with a delicious menu. For breakfast they have hearty classics such as biscuits & gravy, steak & eggs, and French toast. Pulled pork, smoked chicken, ribs, cheeseburgers, beef sandwiches, and rib tips with fries are just a few of the lunch and dinner options. Eat your meal and get ready to be amazed because the totality phase of the eclipse lasts well over 3 minutes in Sikeston. Located at 1609 E Malone Ave, (573) 471-9927

Eat My Catfish in Conway serves up U.S. farm-raised catfish that is never frozen and free-range Arkansas chicken. Alongside their famed catfish dishes, the restaurant features Po'Boy sandwiches and a range of seafood options, including shrimp, crab legs, and crawfish. The sun's temporary disappearance lasts for 3:55 in Conway. Fuel up before the breathtaking event commences at 1:51. Located at 2125 Harkrider St, (501) 588-1867

Feltner's Whatta-Burger in Russellville has a classic drive-in menu. The restaurant is known for its custom handmade burgers, including options such as Whatta-Burger, Whatta-Cheese, Double Meat, Triple Meat, and Mushroom & Swiss Burger, among others. They also have a variety of sandwiches and hot dogs, including Fried Bologna. They have drinks, desserts and fried pies too. Russellville is near the center of the shadow and you'll be amazed by over 4 minutes of totality. Located at 1410 N Arkansas Ave.

Phil's Family Restaurant in Hot Springs offers classic southern comfort food options. Main entrees include country fried steak, hamburger steak, fried ham, fried shrimp, grilled tilapia and catfish. They have rotating daily specials like gumbo, meatloaf, and chicken and dumplings. The total solar eclipse lasts over 3 minutes in this area, so grab some grub and check it out! Located at 2900 Central Ave, (501) 623-8258

The Southerner in Cape Girardeau offers Southern-style cuisine that will leave diners coming back for more. They have appetizers like Dirty Cajun Fries, Three Deviled Eggs Went Down to Georgia, and Crispy Fried Pickle Chips. Signature entrées include Piggy Mac with smoked pulled pork, Not Your Mama's Meatloaf, and Shrimp and Grits. Witness over 4 minutes of sky magic in Cape Girardeau as totality swoops in at 1:58. Be sure to get your food early! Located at 3359 Percy Dr, (573) 708-6530

Yesterdays Restaurant & Bar in Morrilton offers an appetizing and diverse menu. Appetizers include homemade Con Queso & Salsa, Fried Pickles and Fried Green Beans. They serve a variety of wraps, salads, pasta, sandwiches, seafood, burgers, pizzas, and steaks. Morrilton is close to the center line of the eclipse and you will experience over 4 minutes of totality. Have your food ready and enjoy the extraordinary phenomenon. Located at 1502 N Oak St, (501) 354-8821

Eclipse Track Notes - AR - MO - KY

Cities near the northern and southern limits
The cities of Fort Smith, Little Rock, and Arkadelphia lie close to the edge of the total eclipse path. For Fort Smith, to see the total eclipse at all you'll need to be east of State Highway 59 and southeast of Interstate I-49. For Arkadelphia and Little Rock, both cities are within the path of totality but are near the southern edge. If weather permits, for a longer view of totality, consider moving towards Glenwood, Hot Springs, Conway or Morrilton, all of which are closer to the center of the total eclipse.

Mobility
The interstate highways in Arkansas are helpful ways to get around for this eclipse. Interstate I-30 follows the southern limit of the total eclipse in Arkansas and connects Little Rock to Arkadelphia, Texarkana, Sulphur Springs in Texas and Dallas. Interstate I-40 from Little Rock will take you across the eclipse centerline and to Conway, Morrilton, Russellville, and Clarksville. Interstate I-40 east of Little Rock connects with I-55, creating a convenient path north to Cape Girardeau in southern Missouri. US-60 crosses I-55 and runs east and west, connecting Poplar Bluff, Dexter, and Sikeston in Missouri to Paducah in Kentucky.

Hope, AR
Totality Starts at: 1:48pm CDT
Totality Lasts for: 1:41
Totality Ends at: 1:50pm CDT
NWS Office: www.weather.gov/shv
City Website: www.hopearkansas.net

Amigo Juan Mexican Cafe
1300 N Hervey St, (870) 777-0006
Mexican, Hope
Eclipse Day Monday, 11am - 9pm

Big Jake's BBQ
603 W Commerce Blvd, (870) 777-1000
BBQ, Hope
Eclipse Day Monday, 10:30am - 8pm

BigMac's Barbeeque
2703 N Hazel St, (870) 722-8011
BBQ, Hope
Eclipse Day Monday, 11am - 8pm

Dos Loco Gringos
2406 N Hervey St, (870) 777-3377
Mexican, Hope
Eclipse Day Monday, 10:30am - 8pm

El Agaves Mexican Restaurant
501 N Hervey St, (870) 777-4242
Mexican, Hope
Eclipse Day Monday, 11am - 9pm

El Rinconcito De Hope
1100 N Hervey St, (870) 722-8072
Mexican, Hope
Eclipse Day Monday, 11am - 9pm

Sheba's
1013 N Hervey St, (870) 777-6266
American, Hope
Eclipse Day Monday, 6am - 9pm

Tailgaters Burger Co
101 S Main St, (870) 777-4444
American, Hope
Eclipse Day Monday, 11am - 8:30pm

Arkadelphia, AR
Totality Starts at: 1:49pm CDT
Totality Lasts for: 2:13
Totality Ends at: 1:51pm CDT
NWS Office: www.weather.gov/lzk
City Website: arkadelphia.gov

67 Grill
629 Main St, (870) 464-1537
American, Arkadelphia
Eclipse Day Monday, 11am - 10pm

Andy's
2927 Pine St, (870) 246-2714
American, Arkadelphia
Eclipse Day Monday, 5:30am - 9pm

El Parian Mexican Restaurant
202 N 10th St, (870) 245-2546
Mexican, Arkadelphia
Eclipse Day Monday, 11am - 9pm

El Ranchito grill
2805 Pine St, (870) 230-1050
Mexican, Arkadelphia
Eclipse Day Monday, 11am - 9pm

Guti's Mex-Italy Restaurant
2607 Caddo St, (870) 617-7040
Italian, Arkadelphia
Eclipse Day Monday, 11am - 9pm

Great Wall Restaurant
112 Wp Malone Dr Ste A, (870) 246-9800
Chinese, Arkadelphia
Eclipse Day Monday, 10:30am - 8:30pm

Hamburger Barn
2813 Pine St, (870) 246-5556
American, Arkadelphia
Eclipse Day Monday, 11am - 8:30pm

Tasty Hibachi
2607 Caddo St, (870) 617-7063
Japanese, Arkadelphia
Eclipse Day Monday, 11am - 2:30pm, 4–9pm

Walk & Taco
2909 Pine St, (870) 617-7007
Mexican, Arkadelphia
Eclipse Day Monday, 7am - 9pm

Never gaze at the sun without eye protection. Only remove eye protection during 100% totality.

Glenwood, AR
Totality Starts at: 1:48pm CDT
Totality Lasts for: 4:00
Totality Ends at: 1:52pm CDT
NWS Office: www.weather.gov/lzk
City: www.arkansas.com/glenwood

Caddo Cafe
53 Hwy. 70 East Ste. C, (870) 356-2397
American, Glenwood
Eclipse Day Monday, 7am - 9pm

Daddy T's Pizza Shack
244 US-70, (870) 356-3222
Pizza, Glenwood
Eclipse Day Monday, 11am - 9pm

El Diamante Mexican Restaurant
199 US-70, (870) 356-4707
Mexican, Glenwood
Eclipse Day Monday, 10:30am - 9pm

La Oaxaquena
200 US-70, (870) 356-4947
Mexican, Glenwood
Eclipse Day Monday, 10am - 7pm

O K Cafe
233 US-70, (870) 356-3092
American, Glenwood
Eclipse Day Monday, 5am - 3pm

Hot Springs, AR
Totality Starts at: 1:49pm CDT
Totality Lasts for: 3:38
Totality Ends at: 1:53pm CDT
NWS Office: www.weather.gov/lzk
City Website: hotsprings.org

Bubba Brews - On Lake Hamilton
1252 Airport Rd, (501) 547-3186
American, Hot Springs
Eclipse Day Monday, 11am - 9pm

Cafe 1217
1217 Malvern Ave, (501) 318-1094
American, Hot Springs
Eclipse Day Monday, 11am - 6pm

Diablos Tacos & Mezcal
528 Central Ave, (501) 701-4327
Mexican, Hot Springs
Eclipse Day Monday, 11am - 9pm

Jose's Mexican Grill & Cantina
2215 Malvern Ave, (501) 609-9700
Mexican, Hot Springs
Eclipse Day Monday, 11am - 9pm

K and S Golden Raised LLC
3205 Albert Pike Rd, (501) 781-8412
Breakfast, Hot Springs
Eclipse Day Monday, 5am - 1:30pm

La Bodeguita
1313 Central Ave, (501) 620-4967
Mexican, Hot Springs
Eclipse Day Monday, 9:30am - 8:30pm

Mi Pueblito Mexican Restaurant
2070 Airport Rd, (501) 760-4647
Mexican, Hot Springs
Eclipse Day Monday, 11am - 9:30pm

Phil's Family Restaurant
2900 Central Ave, (501) 623-8258
Southern, Hot Springs
Eclipse Day Monday, 7am - 6pm

Quetzal
1105 Albert Pike Rd, (501) 881-4049
Mexican, Hot Springs
Eclipse Day Monday, 10:30am - 9:30pm

Rise & Dine Cafe Inc
1739 Airport Rd suite a, (501) 760-5899
Breakfast, Hot Springs
Eclipse Day Monday, 6am - 2pm

Rocky's Corner
2600 Central Ave, (501) 624-0199
Italian, Hot Springs
Eclipse Day Monday, 11am - 10pm

Rolando's
210 Central Ave, (501) 318-6054
Ecuadorian, Hot Springs Hot Springs
Eclipse Day Monday, 11am - 9pm

Saddlebags Bar & Grill
4977 Albert Pike Rd, (501) 767-2247
American
Eclipse Day Monday, 12pm–12am

Stubby's BBQ
3024 Central Ave, (501) 624-1552
BBQ, Hot Springs
Eclipse Day Monday, 11am - 8pm

Taco Pronto
4350 Central Ave, (501) 525-7309
Tex-Mex, Hot Springs
Eclipse Day Monday, 10:30am - 8pm

Mena, AR
Totality Starts at: 1:47pm CDT
Totality Lasts for: 4:04
Totality Ends at: 1:51pm CDT
NWS Office: www.weather.gov/lzk
City Website: cityofmena.org

Branding Iron BBQ & Steak House
623 Sherwood Ave B, (479) 437-3240
BBQ, Mena
Eclipse Day Monday, 11am - 9pm

Chicollo's Food Emporium
1308 N U.S. 71, (479) 234-8236
American, Mena
Eclipse Day Monday, 11am - 7pm

Chiquita's
703 U. S. Hwy 71, (479) 394-6201
Mexican, Mena
Eclipse Day Monday, 10am - 9pm

La Villa Mexican Restaurant
1100 N U.S. 71, (479) 243-0822
Mexican, Mena
Eclipse Day Monday, 10:30am - 9pm

Myers Cruizzers Drive-In
409 U. S. Hwy 71, (479) 394-5550
American, Mena
Eclipse Day Monday, 8am - 9pm

Rieki Hibachi
410 Sherwood Ave, (870) 490-0351
Sushi, Mena
Eclipse Day Monday, 11am - 2pm, 4–9pm

Skyline Cafe
618 Mena St, (479) 394-5152
American, Mena
Eclipse Day Monday, 7am - 2pm

Waldron, AR
Totality Starts at: 1:48pm CDT
Totality Lasts for: 3:45
Totality Ends at: 1:52pm CDT
NWS Office: www.weather.gov/lzk
City: www.arkansas.com/waldron

Charbroiler Restaurant
N Main St, (479) 637-3163
American, Waldron
Eclipse Day Monday, 11am - 8pm

EL California Mexican Grill
88 Hwy 71 Bypass S, (479) 637-1600
Mexican, Waldron
Eclipse Day Monday, 10:30am - 9pm

El Rincon Mexicano
49 highway 71 bypass south, (479) 637-1109
Mexican, Waldron
Eclipse Day Monday, 11am - 8:30pm

Judy's Drive In
1024 N Main St, (479) 637-0803
American, Waldron
Eclipse Day Monday, 6am - 9pm

Rock Cafe
355 S Main St, (479) 637-2975
American, Waldron
Eclipse Day Monday, 5am - 2pm

Clarksville, AR
Totality Starts at: 1:50pm CDT
Totality Lasts for: 3:30
Totality Ends at: 1:53pm CDT
NWS Office: www.weather.gov/lzk
City Website: www.clarksvillear.gov

China Fun
229 E Market St t3, (479) 754-7500
Chinese, Clarksville
Eclipse Day Monday, 11am - 9pm

Diamond Drive In
1206 W Main St, (479) 754-2160
American, Clarksville
Eclipse Day Monday, 10:30am - 8pm

El Molcajete
101 S Rogers St, (479) 754-2904
Mexican, Clarksville
Eclipse Day Monday, 11am - 9pm

El Parian
711 E Main St, (479) 705-0115
Mexican, Clarksville
Eclipse Day Monday, 11am - 9pm

Kountry Kitchen Grille
2604 W Main St, (479) 754-2611
American, Clarksville
Eclipse Day Monday, 7am - 8pm

La Chiquita Restaurant
133 Taylor Rd, (479) 705-7022
Mexican, Clarksville
Eclipse Day Monday, 10:30am - 9pm

La Michoacana Mexican Grill
1618 W Main St, (479) 754-0081
Mexican, Clarksville
Eclipse Day Monday, 11am - 8:30pm

South Park Restaurant
1157 S Rogers St, (479) 754-8249
American, Clarksville
Eclipse Day Monday, 6am - 9pm

Russellville, AR
Totality Starts at: 1:50pm CDT
Totality Lasts for: 4:10
Totality Ends at: 1:54pm CDT
NWS Office: www.weather.gov/lzk
City: www.russellvillearkansas.org

Fat Daddy's Bar-B-Que of London
7206 US-64 B, (479) 967-1273
BBQ, Russellville
Eclipse Day Monday, 10:30am - 9pm

Feltner's Whatta-Burger
1410 N Arkansas Ave, Russellville, AR 72801
American, Russellville
Eclipse Day Monday, 11am - 8pm

Foodies Restaurant & Grill
2405 E Parkway Dr, (479) 398-2578
Mediterranean, Russellville
Eclipse Day Monday, 11am - 8pm

La Chiquita
1509 E Main St, (479) 890-9402
Mexican, Russellville
Eclipse Day Monday, 11am - 9pm

Las Palmas
615 N Arkansas Ave, (479) 890-2550
Mexican, Russellville
Eclipse Day Monday, 11am - 9pm

Linh Vietnamese Cuisine
624 S Knoxville Ave, (479) 498-6495
Vietnamese, Russellville
Eclipse Day Monday, 10am - 8pm

Old South Restaurant
1330 E Main St, (479) 968-3789
American, Russellville
Eclipse Day Monday, 7am - 9pm

Peg Leg | Handmade Burgers & Fries
2621 W Main St, (479) 692-1869
American, Russellville
Eclipse Day Monday, 10:30am - 8:30pm

Stoby's Restaurant
405 W Parkway Dr, (479) 968-3816
American, Russellville
Eclipse Day Monday, 6am - 9pm

Taco Villa
420 E 4th St, (479) 968-1191
Mexican, Russellville
Eclipse Day Monday, 11am - 9pm

The Gunslingin' Burger
713 E 4th St, (479) 219-3517
American, Russellville
Eclipse Day Monday, 11am - 8pm

Morrilton, AR
Totality Starts at: 1:50pm CDT
Totality Lasts for: 4:12
Totality Ends at: 1:54pm CDT
NWS Office: www.weather.gov/lzk
City Website: cityofmorrilton.com

Blue Diamond Cafe
1800 E Harding St, (501) 354-4253
American, Morrilton
Eclipse Day Monday, 10am - 8pm

Elia's Mexican Grill
1700 University Blvd, (501) 208-5625
Mexican, Morrilton
Eclipse Day Monday, 11am - 8pm

Los Cabos Mexican Grill
912 E Drilling St, (501) 477-2333
Mexican, Morrilton
Eclipse Day Monday, 11am - 9pm

Morrilton Drive Inn Restaurant
1601 N Oak St, (501) 354-8343
American, Morrilton
Eclipse Day Monday, 8am - 9:30pm

Stick's Deli & Diner
510 W Broadway St, (501) 242-1567
Deli, Morrilton
Eclipse Day Monday, 7am - 1pm

Two Guys Hibachi Express
112 N Moose St, (501) 596-1644
Japanese, Morrilton
Eclipse Day Monday, 11am - 2pm, 4–8pm

Yesterday's
1502 N Oak St, (501) 354-8821
American, Morrilton
Eclipse Day Monday, 10:30am - 9pm

Conway, AR
Totality Starts at: 1:51pm CDT
Totality Lasts for: 3:55
Totality Ends at: 1:55pm CDT
NWS Office: www.weather.gov/lzk
conwayarkansas.gov

Bulgogi Korean BBQ
317 Oak St Suite 1, (501) 358-5923
Korean BBQ
Eclipse Day Monday, 11am - 8:30pm

Blaze Pizza
455 Elsinger Blvd, (501) 764-9051
Pizza, Conway
Eclipse Day Monday, 11am - 9pm

Bob's Grill & Cafeteria
1112 Oak St, (501) 329-9760
American, Conway
Eclipse Day Monday, 5am - 2pm

Burgers Pies & Fries
2160 Harkrider St, (501) 358-6110
American, Conway
Eclipse Day Monday, 9am - 9pm

Don Pepe's Gourmet Burritos & Tacos
2225 Prince St, (501) 358-6007
Mexican, Conway
Eclipse Day Monday, 11am - 9pm

Eat My Catfish
2125 Harkrider St, (501) 588-1867
Seafood, Conway
Eclipse Day Monday, 11am - 8pm

Holly's Country Cooking
116 S Harkrider St, (501) 328-9738
American, Conway
Eclipse Day Monday, 11am - 2pm

Hideaway Pizza
1170 S Amity Rd, (501) 697-4444
Pizza, Conway
Eclipse Day Monday, 11am - 9:30pm

Jade China
559 Harkrider St, (501) 329-5121
Chinese, Conway
Eclipse Day Monday, 11am - 9pm

La Huerta Mexican Restaurant
1052 Harrison St, (501) 764-0202
Mexican, Conway
Eclipse Day Monday, 11am - 9pm

Los 3 Potrillos
2490 Sanders Rd, (501) 327-1144
Mexican, Conway
Eclipse Day Monday, 11am - 10pm

MarketPlace Grill
600 Skyline Dr, (501) 336-0011
American, Conway
Eclipse Day Monday, 11am - 8:30pm

Mike's Place
808 Front St, (501) 269-6453
Cajun, Conway
Eclipse Day Monday, 11am - 8:30pm

O'Malley's Irish Grill
803 Harkrider St Suite 12, (501) 504-6949
Irish, Conway
Eclipse Day Monday, 11am - 9pm

Never gaze at the sun without eye protection. Only remove eye protection during 100% totality.

Patron Mexican Grill
1475 Hogan Ln # 123, (501) 328-3265
Mexican, Conway
Eclipse Day Monday, 11am - 9pm

Pho Huyen
1600 Dave Ward Dr ste h, (501) 504-2449
Vietnamese, Conway
Eclipse Day Monday, 11am - 8pm

Sharks Fish & Chicken
2665 N Donaghey Ave, (501) 358-6737
Seafood, Conway
Eclipse Day Monday, 10am - 10pm

Shorty's Bar-B-Que
1101 Harkrider St, (501) 329-9213
BBQ, Conway
Eclipse Day Monday, 10am - 8pm

The Mighty Crab - Conway
2104 Harkrider St Suite 101, (501) 504-2288
Seafood, Conway
Eclipse Day Monday, 11am - 10pm

The Purple Cow Restaurant (Conway)
1055 Steel Ave, (501) 205-4211
American, Conway
Eclipse Day Monday, 11am - 9pm

Wild Ginger Food Truck
580 E Dave Ward Dr, (501) 548-5123
Japanese, Conway
Eclipse Day Monday, 11am - 3pm, 4–8:45pm

ZAZA : Conway
1050 Ellis Ave, (501) 336-9292
Pizza, Conway
Eclipse Day Monday, 11am - 9pm

Clinton, AR
Totality Starts at: 1:51pm CDT
Totality Lasts for: 4:15
Totality Ends at: 1:55pm CDT
NWS Office: www.weather.gov/lzk
City Website: clintonark.com

L'Attitude Bistro
1303 Hwy 65 S, (501) 745-4888
American, Clinton
Eclipse Day Monday, 11am - 8pm

La Rosita Mexican Restaurant
244 US-65, (501) 745-8120
Mexican, Clinton
Eclipse Day Monday, 11am - 9pm

Los Locos Restaurant
933 US-65, (501) 745-2099
Mexican, Clinton
Eclipse Day Monday, 10am - 9pm

Southern Hibachi Express
2143 Hwy 65 S, (501) 253-8108
Asian, Clinton
Eclipse Day Monday, 11am - 3pm, 4–8:30pm

Little Rock/NLR, AR
Totality Starts at: 1:51pm CDT
Totality Lasts for: 2:38
Totality Ends at: 1:54pm CDT
NWS Office: www.weather.gov/lzk
City Website: www.littlerock.gov

Alley Oops
11900 Kanis Rd LR, (501) 221-9400
Southern, Little Rock
Eclipse Day Monday, 11am - 2pm

Bellwood Diner
3815 MacArthur Dr NLR, (501) 753-1012
American, North Little Rock
Eclipse Day Monday, 6am - 2pm

Black Angus
5100 W Markham St LR, (501) 228-7800
Steak, Little Rock
Eclipse Day Monday, 10:30am - 8pm

Bobby's Cafe
18505 MacArthur Dr NLR, (501) 851-7888
American, North Little Rock
Eclipse Day Monday, 6:30am - 2pm

Bobby's Country Cookin'
301 N Shackleford Rd E1 LR, (501) 224-9500
American, Little Rock
Eclipse Day Monday, 10:30am - 2pm

Cantina Cinco De Mayo 1
3 Rahling Cir LR, (501) 821-2740
Mexican, Little Rock
Eclipse Day Monday, 11am - 9pm

Faded Rose Restaurant
1619 Rebsamen Park Rd LR, (501) 663-9734
Seafood, Little Rock
Eclipse Day Monday, 11am - 9pm

Gadwall's Grill
7311 N Hills Blvd #14 NLR, (501) 834-1840
American, North Little Rock
Eclipse Day Monday, 11am - 9pm

Gandolfo's Delicatessen
17801 Chenal Pkwy Suite A LR, (501) 830-4071
Deli, Little Rock
Eclipse Day Monday, 9am - 7pm

Grady's Pizza & Subs
6801 W 12th St LR, (501) 663-1918
Pizza, Little Rock
Eclipse Day Monday, 11am - 8pm

Jacob's Wings & Grill
5200 John F Kennedy Blvd NLR, (501) 508-5783
Wings, North Little Rock
Eclipse Day Monday, 11am - 2:30pm, 4:30–9pm

Kebab House
11321 W Markham St #4 LR, (501) 246-4597
Mediterranean, Little Rock
Eclipse Day Monday, 11am - 8pm

Kosuke Japanese Steakhouse and Sushi
1900 Club Manor Dr Ste 108, (501) 734-8434
Sushi, Maumelle
Eclipse Day Monday, 11am - 2:30pm, 4:30–9pm

Las Palmas
4154 E McCain Blvd NLR, (501) 945-8010
Mexican, North Little Rock
Eclipse Day Monday, 11am - 9:30pm

Leo's Greek Castle
2925 Kavanaugh Blvd LR, (501) 666-7414
American & Mediterranean, Little Rock
Eclipse Day Monday, 7am - 9pm

Little Greek Fresh Grill
11525 Cantrell Rd Suite #905 LR, (501) 223-5300
Greek, Little Rock
Eclipse Day Monday, 11am - 8pm

Maryam's Grill Mediterranean Restaurant
323 Center St LR, (501) 374-2633
Mediterranean, Little Rock
Eclipse Day Monday, 8am - 2pm

Morgan's Kitchen & Deli
2801 W 7th St LR, (501) 515-4518
American, Little Rock
Eclipse Day Monday, 10:30am - 6:30pm

Ocean's At Arthurs
16100 Chenal Pkwy LR, (501) 821-1828
Seafood, Little Rock
Eclipse Day Monday, 11am - 2pm, 4:30–9:30pm

Rosalinda's Restaurant
900 W 35th St NLR, (501) 771-5559
Honduran, North Little Rock
Eclipse Day Monday, 10am - 9pm

Sauced Bar and Oven
11121 N Rodney Parham Rd LR, (501) 353-1534
Pizza, Little Rock
Eclipse Day Monday, 11am - 9pm

Señor Tequila
4304 Camp Robinson Rd NLR, (501) 791-3888
Mexican, North Little Rock
Eclipse Day Monday, 11am - 10pm

Sim's Bar-B-Que
1307 John Barrow Rd A LR, (501) 224-2057
BBQ, Little Rock
Eclipse Day Monday, 11am - 9pm

Taqueria Azteca Original Mexican Food
13503 Crystal Hill Rd NLR, (501) 800-1088
Mexican, North Little Rock
Eclipse Day Monday, 10am - 9pm

Taco Mexicano
6012 Crystal Hill Rd NLR, (501) 554-4327
Mexican, North Little Rock
Eclipse Day Monday, 10am - 5pm

Tacos Atilano - NLR
3501 Pike Ave, (870) 877-1152
Mexican, North Little Rock
Eclipse Day Monday, 11am - 11pm

The Pantry Restaurant
11401 N Rodney Parham Rd LR, (501) 353-1875
European, Little Rock
Eclipse Day Monday, 11am - 10pm

The Pizza Cafe West
14710 Cantrell Rd LR, (501) 868-2600
Pizza, Little Rock
Eclipse Day Monday, 11am - 9pm

Never gaze at the sun without eye protection. Only remove eye protection during 100% totality.

Mountain Home, AR
Totality Starts at: 1:53pm CDT
Totality Lasts for: 3:10
Totality Ends at: 1:56pm CDT
NWS Office: www.weather.gov/lzk
City: www.cityofmountainhome.com

Arena Sports Grill
1041 Highland Cir, (870) 425-5355
American, Mountain Home
Eclipse Day Monday, 11am - 10pm

Brenda's Cafe
3555 US-62, (870) 492-5955
American, Mountain Home
Eclipse Day Monday, 5am - 2pm

Buncles Brickoven & Brews
1406 Hwy 62 E, (870) 425-2337
Pizza / American, Mountain Home
Eclipse Day Monday, 11am - 9pm

Dusit Thai Cuisine
343 Hwy 62 E, (870) 425-7272
Thai, Mountain Home
Eclipse Day Monday, 10:30am - 2:30pm, 4–8pm

El Charro Mexican Restaurant
337 Hwy 62 E, (870) 425-7877
Mexican, Mountain Home
Eclipse Day Monday, 10:45am - 9pm

Holy Smokes BBQ
400 AR-201, (870) 425-8080
BBQ, Mountain Home
Eclipse Day Monday, 11am - 8pm

LaPaloma Mountain Home
1014 Highland Cir, (870) 424-2310
Mexican, Mountain Home
Eclipse Day Monday, 11am - 9pm

Mel's Diner
860 Hwy 62 E, (870) 424-0469
American, Mountain Home
Eclipse Day Monday, 7am - 2pm

O. M. Greek
912 Hwy 62 E, (870) 701-5207
Greek, Mountain Home
Eclipse Day Monday, 10:30am - 4pm

Salsa's Grill
965 Hwy 62 E, (870) 424-2690
Mexican, Mountain Home
Eclipse Day Monday, 11am - 9:30pm

Skipper's Restaurant
711 AR-5, (870) 508-4574
Seafood, Mountain Home
Eclipse Day Monday, 7am - 2pm

The Back Forty
1400 Hwy 62 E, (870) 425-7170
American, Mountain Home
Eclipse Day Monday, 11am - 10pm

The Cozy Kitchen
1611 US-62 BUS, (870) 425-0500
Breakfast, Mountain Home
Eclipse Day Monday, 6am - 2pm

Mountain View, AR
Totality Starts at: 1:52pm CDT
Totality Lasts for: 4:14
Totality Ends at: 1:56pm CDT
NWS Office: www.weather.gov/lzk
City Website: www.cityofmtnview.org

Bushel And A Peck Cafe
201 W Main St, (870) 269-3365
American, Mountain View
Eclipse Day Monday, 10am - 4pm

Kin Folk's BBQ
118 Howard Ave, (870) 269-9188
BBQ, Mountain View
Eclipse Day Monday, 11am - 7:30pm

Krispy House
706 E Main St, (870) 269-3144
American, Mountain View
Eclipse Day Monday, 10am - 6pm

Los Locos Mexican Restaurant
204 B, 204 Sylamore Ave A, (870) 269-5404
Mexican, Mountain View
Eclipse Day Monday, 10am - 9pm

Mi Pueblito 4 Mountain View
103 N Peabody Ave A, (870) 269-6400
Mexican, Mountain View
Eclipse Day Monday, 11am - 9pm

Wing Shack-Cheeseburger Grill
305 Sylamore Ave, (870) 269-9464
American, Mountain View
Eclipse Day Monday, 11am - 9pm

Batesville, AR
Totality Starts at: 1:53pm CDT
Totality Lasts for: 4:03
Totality Ends at: 1:57pm CDT
NWS Office: www.weather.gov/lzk
City: www.cityofbatesville.com

China King Super Buffet
1238 N St Louis, (870) 612-1800
Chinese, Batesville
Eclipse Day Monday, 11am - 9pm

Cowboy's Barbecue
4101 Harrison St, (870) 698-8500
BBQ, Batesville
Eclipse Day Monday, 11am - 8pm

Dora Express Japanese Food
2370 Harrison St, (870) 834-7777
Japanese, Batesville
Eclipse Day Monday, 11am - 2pm, 4–8:30pm

El Palenque Mexican Restaurant
2518 Harrison St, (870) 793-8331
Mexican, Batesville
Eclipse Day Monday, 11am - 8pm

Las Playitas
350 Harrison St, (870) 569-4339
Mexican, Batesville
Eclipse Day Monday, 11am - 9:30pm

MI Ranchito III
5 Eagle Mountain Blvd, (870) 793-8227
Mexican, Batesville
Eclipse Day Monday, 11am - 9pm

Morningside Coffee House
616 Harrison St, (870) 793-3335
Coffee Shop, Batesville
Eclipse Day Monday, 7am - 2pm

Natalie's Cafe & Catering
3050 Harrison St, (870) 698-0200
American, Batesville
Eclipse Day Monday, 11am - 3pm

Nova Joe's
2100 Harrison St, (870) 569-5118
Coffee Shop, Batesville
Eclipse Day Monday, 5am - 9pm

South Side Grill
2121 Batesville Blvd, (870) 251-2229
Southern, Batesville
Eclipse Day Monday, 10:30am - 2pm

Spah Grill Batesville
763 Batesville Blvd, (870) 612-5600
American, Batesville
Eclipse Day Monday, 6am - 2:30pm

Stella's Brick Oven Pizzeria and Bistro
250 E Main St, (870) 805-3045
Pizza, Batesville
Eclipse Day Monday, 11am - 2pm, 4–8pm

Whistle Stop Cafe
300 Lawrence St, (870) 793-0997
American, Batesville
Eclipse Day Monday, 6am - 7pm

Ash Flat, AR
Totality Starts at: 1:53pm CDT
Totality Lasts for: 4:12
Totality Ends at: 1:57pm CDT
NWS Office: www.weather.gov/lzk
City Website: www.ash-flat.com

Artisan Grill
190 Hwy 62 412, (870) 751-2034
American
Eclipse Day Monday, 11am - 9pm

Chokdee Thai Kitchen
595 Ash Flat Dr, (870) 994-0907
Thai, Ash Flat
Eclipse Day Monday, 11am - 8pm

El Palenque Mexican Cuisine
412 US-62, (870) 994-7700
Mexican, Ash Flat
Eclipse Day Monday, 11am - 9pm

Patio Lino
W 298 US-62, (870) 994-2904
Mexican, Ash Flat
Eclipse Day Monday, 11am - 8pm

Never gaze at the sun without eye protection. Only remove eye protection during 100% totality.

Missouri

Poplar Bluff, MO
Totality Starts at: 1:56pm CDT
Totality Lasts for: 4:08
Totality Ends at: 2:00pm CDT
NWS Office: www.weather.gov/pah
City Website: www.poplarbluff-mo.gov

bread+butter
2586 N Westwood Blvd, (573) 785-8500
American, Poplar Bluff
Eclipse Day Monday, 10:30am - 8pm

Casa Grande
2027 N Westwood Blvd, (573) 727-9585
Mexican, Poplar Bluff
Eclipse Day Monday, 11am - 9pm

Castello's Restaurant & Catering LLC
2775 N Westwood Blvd, (573) 712-9010
Italian, Poplar Bluff
Eclipse Day Monday, 5:30am - 9pm

China Wok
368 N 6th St, (573) 686-8821
Chinese, Poplar Bluff
Eclipse Day Monday, 10:30am - 9:30pm

Dexter Bar-B-Que
101 N Westwood Blvd, (573) 785-1900
BBQ, Poplar Bluff
Eclipse Day Monday, 11am - 8pm

Earl's Diner
N F St, (573) 727-9844
American, Poplar Bluff
Eclipse Day Monday, 5am - 2pm

El Acapulco Authentic Mexican
2260 N Westwood Blvd, (573) 776-7000
Mexican, Poplar Bluff
Eclipse Day Monday, 11am - 9pm

Fuji Japanese Cuisine
1385 N Westwood Blvd, (573) 609-2246
Japanese, Poplar Bluff
Eclipse Day Monday, 11am - 2:30pm, 4:30–9pm

Hayden Drive-In
807 W Maud St, (573) 785-4705
BBQ, Poplar Bluff
Eclipse Day Monday, 8am - 3pm

Jen's Diner
3933 S Westwood Blvd, (573) 712-2155
American, Poplar Bluff
Eclipse Day Monday, 5am - 2pm

Lemonade House Grille
2789 Tucker Rd, (573) 776-8053
American, Poplar Bluff
Eclipse Day Monday, 10:30am - 8pm

Las Margaritas
2144 N Westwood Blvd, (573) 686-3246
Mexican, Poplar Bluff
Eclipse Day Monday, 11am - 10pm

Maya's Mexican Restaurant
940 S Westwood Blvd, (573) 785-7966
Mexican, Poplar Bluff
Eclipse Day Monday, 11am - 10pm

Myrtle's Place
109 N Broadway St, (573) 785-9203
American, Poplar Bluff
Eclipse Day Monday, 5am - 3pm

Taco Taco Fresh Mex
1405 N Westwood Blvd, (573) 686-8889
Mexican, Poplar Bluff
Eclipse Day Monday, 11am - 9pm

Tios Bar & Grill
1135 Herschel Bess Blvd, (573) 778-0217
Mexican, Poplar Bluff
Eclipse Day Monday, 11am - 9pm

Dexter, MO
Totality Starts at: 1:57pm CDT
Totality Lasts for: 3:53
Totality Ends at: 2:01pm CDT
NWS Office: www.weather.gov/pah
City Website: visitdexter.com

Airways Cafe
Harry Bennett Dr, (573) 624-4377
American, Dexter
Eclipse Day Monday, 6am - 2pm

Corner Stop Cafe
5 S Locust St, (573) 614-5484
American, Dexter
Eclipse Day Monday, 8am - 3pm

Dexter Bar-B-Q - Dexter
1411 W Business U.S. 60, (573) 624-8810
BBQ, Dexter
Eclipse Day Monday, 11am - 8pm

Dexter Queen
1109 W Business U.S. 60, (573) 624-2330
BBQ, Dexter
Eclipse Day Monday, 11am - 9pm

El Cabrito Mexican Restaurant
208 W. Business U.S. Highway 60, (573) 614-4205
Mexican, Dexter
Eclipse Day Monday, 11am - 9pm

El Tapatio of Dexter
1612 W Business U.S. 60, (573) 624-8662
Mexican, Dexter
Eclipse Day Monday, 11am - 10pm

Hickory Log Restaurant
1314 W Business U.S. 60, (573) 624-4950
BBQ, Dexter
Eclipse Day Monday, 11am - 9pm

Peking Chinese Restaurant LLC
1622 W Business U.S. 60, (573) 624-6888
Chinese, Dexter
Eclipse Day Monday, 10:30am - 10pm

Sikeston, MO
Totality Starts at: 1:58pm CDT
Totality Lasts for: 3:31
Totality Ends at: 2:01pm CDT
NWS Office: www.weather.gov/pah
City Website: www.sikeston.org

Bo's Breakfast and Bar-B-Q
1609 E Malone Ave, (573) 471-9927
BBQ, Sikeston
Eclipse Day Monday, 6am - 2pm

Dexter Bar-B-Q - Sikeston
124 N Main St, (573) 471-6676
BBQ, Sikeston
Eclipse Day Monday, 11am - 8pm

El Bracero of Sikeston
2612 E Malone Ave, (573) 472-0616
Mexican, Sikeston
Eclipse Day Monday, 11am - 10pm

El Cerrito Mexican Restaurant
1103 E Malone Ave, (573) 621-5122
Mexican, Sikeston
Eclipse Day Monday, 10am - 9pm

El Nopalito Grill
2600 E Malone Ave, (573) 475-7099
Mexican, Sikeston
Eclipse Day Monday, 10am - 10pm

El Tapatio of Sikeston
2113 E Malone Ave, (573) 472-3888
Mexican, Sikeston
Eclipse Day Monday, 11am - 10pm

Grecian Steak House
531 Greer Ave, (573) 471-6877
Steak, Sikeston
Eclipse Day Monday, 11am - 9pm

Jay's Krispy Fried Chicken
218 N Main St, (573) 471-8472
Chicken, Sikeston
Eclipse Day Monday, 6am - 9pm

La Ruleta Mexican Restaurant
1189 S Main St suite c, (573) 471-7789
Mexican, Sikeston
Eclipse Day Monday, 11am - 10pm

Never gaze at the sun without eye protection. Only remove eye protection during 100% totality.

Lambert's Cafe
2305 E Malone Ave, (573) 471-4261
American, Sikeston
Eclipse Day Monday, 10:30am - 8pm

Peking
929 S Kingshighway, (573) 472-6190
Chinese, Sikeston
Eclipse Day Monday, 10:30am - 10pm

Susie's Bake Shop & Restaurant
112 E Center St, (573) 471-8550
American, Sikeston
Eclipse Day Monday, 8am - 5pm

Watami
1229 Hennings, (573) 621-5022
Japanese, Sikeston
Eclipse Day Monday, 11am - 9pm

Cape Girardeau, MO
Totality Starts at: 1:58pm CDT
Totality Lasts for: 4:06
Totality Ends at: 2:02pm CDT
NWS Office: www.weather.gov/pah
City: www.cityofcapegirardeau.org

Abelardo's Mexican Fresh
1740 Broadway St, (573) 803-1398
Mexican, Cape Girardeau
Eclipse Day Monday, 7am - 12am

Bandanas Bar-B-Q Cape Girardeau
156 Vantage Dr, (573) 519-3009
BBQ, Cape Girardeau
Eclipse Day Monday, 11am - 9pm

Bella Italia
20 N Spanish St, (573) 332-7800
Italian, Cape Girardeau
Eclipse Day Monday, 11am - 9pm

BG's Deli
205 S Plaza Way, (573) 335-8860
Deli, Cape Girardeau
Eclipse Day Monday, 11am - 9pm

Burrito-Ville
913 Broadway St, (573) 334-4068
Mexican, Cape Girardeau
Eclipse Day Monday, 10:30am - 9pm

Cafe & Me
820 N Sprigg St #5, (573) 334-2777
Asian, Cape Girardeau
Eclipse Day Monday, 11am - 3pm, 5–8pm

Don Carlos Authentic Mexican And Taqueria
221 S Broadview St, (573) 332-8226
Mexican, Cape Girardeau
Eclipse Day Monday, 11am - 9pm

El Sol
1105 Broadway St, (573) 803-2775
Mexican, Cape Girardeau
Eclipse Day Monday, 11am - 8pm

El Torero Mexican Grill
2120 William St, (573) 339-2040
Mexican, Cape Girardeau
Eclipse Day Monday, 11am - 9:30pm

Krabby Daddy's Cape Girardeau
841 N Kingshighway St, (573) 803-2229
Seafood, Cape Girardeau
Eclipse Day Monday, 11am - 8pm

La Luna
97 N Kingshighway St Ste 10, (573) 388-4818
Mexican, Cape Girardeau
Eclipse Day Monday, 11am - 9pm

Little Kitchen
1036 N Sprigg St, (573) 803-0955
Chinese, Cape Girardeau
Eclipse Day Monday, 11am - 9pm

Logan's Roadhouse
3012 William St, (573) 651-4142
American, Cape Girardeau
Eclipse Day Monday, 11am - 10pm

Muy Bueno Mexican Restaurant
1751 Independence St, (573) 335-6413
Mexican, Cape Girardeau
Eclipse Day Monday, 11am - 9pm

Penn Station East Coast Subs
127 Siemers Dr, (573) 332-0056
American, Cape Girardeau
Eclipse Day Monday, 10:30am - 10pm

Rice Noodle Cai
1017 Independence St, (573) 339-8753
Chinese, Cape Girardeau
Eclipse Day Monday, 11am - 8:30pm

Sand's Pancake House
602 Morgan Oak St, (573) 335-0420
Pancakes, Cape Girardeau
Eclipse Day Monday, 5am - 2pm

Sedona Bistro
1812 Carondelet Dr #101, (573) 803-5000
American, Cape Girardeau
Eclipse Day Monday, 11am - 8pm

Shogun Japanese Sushi and Hibachi
161 West Dr, (573) 335-3090
Japanese, Cape Girardeau
Eclipse Day Monday, 11am - 2:30pm, 4:30–9:30

The Southerner
3359 Percy Dr, (573) 708-6530
American, Cape Girardeau
Eclipse Day Monday, 11am - 9pm

Kentucky

Paducah, KY
Totality Starts at: 2:00pm CDT
Totality Lasts for: 2:00 Northwest
Totality Ends at: 2:00pm CDT
NWS Office: www.weather.gov/pah
City Website: www.paducahky.gov

Artisan Kitchen
1704 Broadway St, (270) 538-0250
American, Paducah
Eclipse Day Monday, 8am - 4pm

Bob's Drive-In Restaurant
2429 Bridge St, (270) 443-6493
American, Paducah
Eclipse Day Monday, 10am - 8pm

Burger Theory
First Level, 600 N 4th St, (270) 366-7614
American, Paducah
Eclipse Day Monday, 7am - 2pm, 5–11pm

El SAZON
2602 Jackson St, (270) 558-4756
Mexican, Paducah
Eclipse Day Monday, 10:30am - 7pm

El Torito Mexican Grill
5450 Cairo Rd, (270) 408-1815
Mexican, Paducah
Eclipse Day Monday, 11am - 10pm

Gold Rush Cafe
400 Broadway St, (270) 443-4422
American, Paducah
Eclipse Day Monday, 7am - 2pm

Happy's Chili Parlor
514 N 12th St, (270) 709-3027
Soul Food, Paducah
Eclipse Day Monday, 11am - 7pm

Hokkaido Grill and Sushi
156 Bleich Rd, (270) 408-6000
Sushi, Paducah
Eclipse Day Monday, 11am - 8:30pm

Hugo's Mexican Restaurant
5101 Hinkleville Rd #506, (270) 443-1020
Mexican, Paducah
Eclipse Day Monday, 11am - 9pm

J. Bella's Pizzeria
2406 New Holt Rd, (270) 554-1123
Pizza, Paducah
Eclipse Day Monday, 11am - 7pm

Lone Oak Little Castle
3460 Lone Oak Rd, (270) 534-9050
Breakfast, Paducah
Eclipse Day Monday, 5am - 7pm

Loopy Larry's
3160 Parisa Dr, (270) 201-2135
American, Paducah
Eclipse Day Monday, 10:30am - 4pm

Los Amigos Mexican Restaurant
5135 Hinkleville Rd, (270) 575-3285
Mexican, Paducah
Eclipse Day Monday, 11am - 9:30pm

Mel's Diner
409 Bleich Rd, (270) 554-4034
American, Paducah
Eclipse Day Monday, 6am - 2pm

Never gaze at the sun without eye protection. Only remove eye protection during 100% totality.

Over/Under
314 Broadway St, (270) 201-2047
American, Paducah
Eclipse Day Monday, 11am - 10pm

Paducah grill
2516 Bridge St, (270) 933-0655
American, Paducah
Eclipse Day Monday, 10:30am - 8pm

Parker's Drive-in and Catering
2921 Lone Oak Rd, (270) 554-7602
American, Paducah
Eclipse Day Monday, 10am - 7pm

Pizza By the Pound
600 N 32nd St, (270) 442-2063
Pizza, Paducah
Eclipse Day Monday, 11am - 9pm

Morganfield, KY
Totality Starts at: 2:01pm CDT
Totality Lasts for: 2:43
Totality Ends at: 2:04pm CDT
NWS Office: www.weather.gov/pah
City Website: morganfield.ky.gov

B & H Restaurant
112 N Airline Rd, (270) 389-0258
American, Morganfield
Eclipse Day Monday, Open 24 hours

El Mexicano
1019 US-60, (270) 389-9603
Mexican, Morganfield
Eclipse Day Monday, 11am - 8:30pm

The Feed Mill Restaurant
3541 US-60 E, (270) 389-0047
American, Morganfield
Eclipse Day Monday, 10:30am - 9pm

Henderson, KY
Totality Starts at: 2:02pm CDT
Totality Lasts for: 2:32
Totality Ends at: 2:05pm CDT
NWS Office: www.weather.gov/pah
City Website: www.hendersonky.gov

Agaves Mexican Grill
2003 Stapp Dr, (270) 957-5028
Mexican, Henderson
Eclipse Day Monday, 11am - 9pm

Bangies Cafe
2036 Madison St, (270) 869-8999
American, Henderson
Eclipse Day Monday, 5:30am - 1pm

Cancun Mexican Restaurant
341 S Green St, (270) 826-0067
Mexican, Henderson
Eclipse Day Monday, 11am - 9pm

Metzger's Tavern
1000 Powell St, (270) 826-9461
American, Henderson
Eclipse Day Monday, 9am - 8pm

Mama's Pizza
1526 N Green St, (270) 830-0101
Pizza, Henderson
Eclipse Day Monday, 11am - 10pm

Mister B's Pizza & Wings
2611 US Hwy 41, (270) 826-9999
Pizza, Henderson
Eclipse Day Monday, 11am - 9:30pm

Rockhouse On The River
212 N Water St, (270) 212-1400
Pizza, Henderson
Eclipse Day Monday, 11am - 9pm

Tacoholics Kitchen
122 1st St, (270) 957-5001
Mexican, Henderson
Eclipse Day Monday, 11am - 9pm

Featured Parks and Attractions

Crater of Diamonds State Park, located in Murfreesboro, Arkansas, is the only diamond-producing site in the world open to the public. Visitors can try their luck at diamond hunting in the 37-acre plowed field, which is the eroded surface of an ancient volcanic crater where diamonds are still regularly found. Any gems or minerals found are yours to keep, making it a one-of-a-kind treasure hunting experience. Beyond the diamond field, the park offers other recreational opportunities including hiking, bird watching, camping, and a water park for cooling off in the summer months. Watch for the "**diamond ring effect**" during the eclipse!

Elephant Rocks State Park, in the Saint Francois Mountains of Missouri, is a geological reserve and public recreation area. The park is named for its signature feature - a string of massive granite boulders that resemble a train of pink elephants. These "elephant rocks" are remnants of weathered Graniteville Granite, a substance that formed 1.4 billion years ago. The rocky surface of the moon covers the sun for almost two and a half minutes during the eclipse, don't miss it!

Hot Springs National Park, is a unique blend of natural and urban experiences. Called "The American Spa," this national park is centered around the hot springs, which have been renowned for their healing properties. The park offers the opportunity to bathe in the thermal waters at the historic Bathhouse Row. The park also features over 26 miles of hiking trails through forests and scenic mountain views, picnic areas, and campgrounds. Three minutes and forty-one seconds of darkness await you here on eclipse day.

John James Audubon State Park, in Kentucky, spans approximately 700 acres, consisting of hilly forests and old-growth trees. The park is home to the internationally acclaimed Audubon Museum, which showcases a multitude of valuable Audubon art pieces, in addition to housing a gift shop and nature center. Additional attractions include a golf course, cottages, a 69-acre campground, and miles of forest hiking trails. Enjoy the park and experience over two and a half minutes of totality here on eclipse day.

Mammoth Spring State Park, an acclaimed National Natural Landmark, is home to one of the world's largest springs. With a remarkable nine million gallons of water flowing hourly, it forms a 10-acre lake. Visitors can explore the remnants of a mill and hydroelectric plant that form part of the park's rich history. The park also showcases an 1886 Frisco train depot and museum. This park is very close to the center line of the total solar eclipse and visitors here will experience four minutes of totality.

Mount Magazine State Park is home to the highest point in Arkansas. There are opportunities for ATV riding, backpacking, hang gliding, mountain biking, and rock climbing. Wildlife sightings are common, with animals such as black bears, whitetail deer, bobcats, and coyotes calling the park home. The park also offers camping, picnic areas, hiking and horseback riding trails, and fishing lakes. Totality lasts for almost four minutes at this state park.

The parks and attractions section in this book are organized by cities, which are highlighted in bold. Under each city, you will find a curated list of locations presented in alphabetical order. These locations may be within the city or situated within a 30-50 mile radius, encompassing the surrounding region. This method of organization allows for easy navigation and planning, ensuring you can make the most of your solar eclipse experience.

Never gaze at the sun without eye protection. Only remove eye protection during 100% totality.

Parks and Attractions

Hope, AR
Totality Starts at: 1:48pm CDT
Totality Lasts for: 1:41
Totality Ends at: 1:50pm CDT
NWS Office: www.weather.gov/shv
City Website: www.hopearkansas.net

Cottonshed Park (AR)
Ashdown, AR 71822
recreation.gov
Totality Lasts for: 3:33

Fair Park
800 S Mockingbird Ln
Hope, AR 71801
www.hopearkansas.net

Historic Washington State Park
103 Franklin St, Washington, AR 71862
historicwashingtonstatepark.com
Totality Lasts for: 2:47

Klipsch Museum of Audio History
136 Hempstead 278 Rd, Hope, AR 71801
www.klipschmuseum.org
Totality Lasts for: 2:21

Millwood State Park
1564 AR-32, Ashdown, AR 71822
arkansasstateparks.com/parks/millwood-state-park
Totality Lasts for: 3:16

Northside Park
1709 N Spruce St
Hope, AR 71801
www.hopearkansas.net

President Clinton Birthplace
117 S Hervey St
Hope, AR 71801
www.nps.gov/wicl/index.htm

Arkadelphia, AR
Totality Starts at: 1:49pm CDT
Totality Lasts for: 2:13
Totality Ends at: 1:51pm CDT
NWS Office: www.weather.gov/lzk
City Website: arkadelphia.gov

Arkadelphia Aquatic Park
2575 Twin Rivers Dr
Arkadelphia, AR 71923
arkadelphia.gov

DeGray Lake State Park Resort
2027 State Park Entrance Rd, Bismarck
degray.com
Totality Lasts for: 3:01

DeGray Lake Sunset Trail
Co Rd 433, Caddo Valley, AR 71923
Totality Starts at: 1:49pm CDT
Totality Lasts for: 2:38

River Park
Arkadelphia, AR 71923
Totality Starts at: 1:49pm CDT
Totality Lasts for: 2:06

Glenwood, AR
Totality Starts at: 1:48pm CDT
Totality Lasts for: 4:00
Totality Ends at: 1:52pm CDT
NWS Office: www.weather.gov/lzk
www.arkansas.com/glenwood

Crater of Diamonds State Park
209 State Park Rd, Murfreesboro, AR 71958
arkansasstateparks.com
Totality Lasts for: 3:40

Daisy State Park
103 E Park Rd, Kirby, AR 71950
arkansasstateparks.com/parks/daisy-state-park
Totality Lasts for: 4:05

Glenwood Access Caddo River
903 3rd St, Glenwood, AR 71943
Totality Starts at: 1:48pm CDT
Totality Lasts for: 4:00

John Benjamin Public Fishing Lake
25-53 Industial Park Ln, Glenwood, AR 71943
Totality Starts at: 1:48pm CDT
Totality Lasts for: 3:59

Hot Springs, AR
Totality Starts at: 1:49pm CDT
Totality Lasts for: 3:38
Totality Ends at: 1:53pm CDT
NWS Office: www.weather.gov/lzk
City Website: hotsprings.org

Arkansas Alligator Farm & Petting Zoo
847 Whittington Ave
Hot Springs, AR 71901
alligatorfarmzoo.com

Baseball Trail Park
1215 Whittington Ave
Hot Springs, AR 71901
www.cityhs.net/210/City-Parks

Cedar Glades Park
461 Wildcat Rd, Hot Springs, AR 71913
www.garlandcounty.org/263/Cedar-Glades-Park
Totality Lasts for: 3:46

Entergy Park
530 Lakepark Dr
Hot Springs, AR 71913
Totality Lasts for: 3:23

Family Fun Park
3500 Central Ave
Hot Springs, AR 71913
hotspringsfamilyfunpark.com

Family Park
215 Family Park Rd
Hot Springs, AR 71913
www.cityhs.net/210/City-Parks

Fordyce Bathhouse Visitor Center And Museum
369 Central Ave
Hot Springs, AR 71901
www.nps.gov/hosp/index.htm

Funtrackers Family Fun Park
2614 Albert Pike Rd
Hot Springs, AR 71913
funtrackersfamilypark.com

Gangster Museum of America
510 Central Ave, Hot Springs, AR 71901
Museum
www.tgmoa.com

Garvan Woodland Gardens
550 Arkridge Rd, Hot Springs, AR 71913
Botanical Garden
www.garvangardens.org

Goat Rock View Point / Trail
Hot Springs National Park, AR 71901
www.nps.gov/places/goat-rock-view-point.htm
Totality Lasts for: 3:41

Hollywood Park
400-425 Hollywood Ave
Hot Springs, AR 71901
www.cityhs.net/210/City-Parks

Hot Springs Mountain Tower
401 Hot Springs Mountain Dr
Hot Springs, AR 71901
hotspringstower.com

Hot Springs National Park
Hot Springs, AR 71901
nps.gov/hosp/
Totality Lasts for: 3:41

Hot Springs National Park KOA Holiday
838 McClendon Rd
Hot Springs, AR 71901
koa.com/campgrounds/hot-springs-national-park/

Josephine Tussaud Wax Museum
250 Central Ave
Hot Springs, AR 71901
seehotsprings.com

Lake Catherine State Park
1200 Catherine Park Rd, Hot Springs
arkansasstateparks.com
Totality Lasts for: 3:05

Lake Ouachita State Park
5451 Mountain Pine Rd, Mountain Pine
arkansasstateparks.com
Totality Lasts for: 4:01

Never gaze at the sun without eye protection. Only remove eye protection during 100% totality.

Mid-America Science Museum
500 Mid America Blvd
Hot Springs, AR 71913
midamericamuseum.org

Pirate's Cove Adventure Golf
4612 Central Ave
Hot Springs, AR 71913
www.piratescove.net/hot-springs

The Galaxy Connection
536 Ouachita Ave
Hot Springs National Park, AR 71901
www.thegalaxyconnection.com

Tiny Town
374 Whittington Ave
Hot Springs, AR 71901
www.tinytowntrains.com

Mena, AR
Totality Starts at: 1:47pm CDT
Totality Lasts for: 4:04
Totality Ends at: 1:51pm CDT
NWS Office: www.weather.gov/lzk
City Website: cityofmena.org

Cossatot River State Park - Natural Area
1980 US-278, Wickes, AR 71973
Totality Starts at: 1:47pm CDT
Totality Lasts for: 4:18

Queen Wilhelmina State Park
3877 AR-88, Mena, AR 71953
Totality Starts at: 1:47pm CDT
Totality Lasts for: 3:46

Waldron, AR
Totality Starts at: 1:48pm CDT
Totality Lasts for: 3:45
Totality Ends at: 1:52pm CDT
NWS Office: www.weather.gov/lzk
www.arkansas.com/waldron

Big Pine RV Park
2137 N Main St
Waldron, AR 72958
www.bigpinervparkar.com

Davison Pool
Waldron, AR 72958
Totality Starts at: 1:48pm CDT
Totality Lasts for: 3:45

Little Pines Recreation Area
Waldron, AR 72958
www.fs.usda.gov
Totality Lasts for: 3:45

Truman Baker Lake Park
Waldron, AR 72958
Totality Starts at: 1:48pm CDT
Totality Lasts for: 3:45

Waldron City Park
Waldron, AR 72958
Totality Starts at: 1:48pm CDT
Totality Lasts for: 3:45

Clarksville, AR
Totality Starts at: 1:50pm CDT
Totality Lasts for: 3:30
Totality Ends at: 1:53pm CDT
NWS Office: www.weather.gov/lzk
City Website: www.clarksvillear.gov

Clarksville Aquatic Center
1611 W Oakland St
Clarksville, AR 72830
www.clarksvillear.gov

Cline Park
208 Meadow Pl, Clarksville, AR 72830
Totality Starts at: 1:50pm CDT
Totality Lasts for: 3:30

Spadra Park
Jamestown Rd
Clarksville, AR 72830
allparx.com/places/spadra-park/

Russellville, AR
Totality Starts at: 1:50pm CDT
Totality Lasts for: 4:10
Totality Ends at: 1:54pm CDT
NWS Office: www.weather.gov/lzk
www.russellvillearkansas.org

Bona Dea Trails & Sanctuary
Scenic 7 Byway
Russellville, AR 72801
www.arkansas.com/russellville

Dam Site Public Use Area
1469 Lock and Dam Rd, Russellville
Totality Starts at: 1:50pm CDT
Totality Lasts for: 4:10

Lake Dardanelle State Park
100 State Park Drive, Russellville
Totality Starts at: 1:49pm CDT
Totality Lasts for: 4:09

Mount Magazine State Park
577 Lodge Dr, Paris, AR 72855
Totality Starts at: 1:49pm CDT
Totality Lasts for: 3:50

Mount Nebo State Park
16728 State Hwy 155, Dardanelle
Totality Starts at: 1:49pm CDT
Totality Lasts for: 4:09

Old Post Park Campground
Old Post Park Rd
Russellville, AR 72802
recreation.gov/camping/campgrounds/232662

Pleasant View Park
3595 N Arkansas Ave, Russellville
Totality Starts at: 1:50pm CDT
Totality Lasts for: 4:07

Potts Inn Museum
E Ash St
Pottsville, AR 72858
www.pottsinnmuseum.weebly.com

Riverview Recreation Area
Dardanelle, AR 72834
Totality Starts at: 1:49pm CDT
Totality Lasts for: 4:11

Russellville Aquatic Center
1300 N Phoenix Ave
Russellville, AR 72801
russellvillearkansas.org

Russellville City Recreation
1000 E Parkway Dr
Russellville, AR 72801
russellvillearkansas.org

Washburn Park
72801, 1185 Lake Front Dr, Russellville
Totality Starts at: 1:50pm CDT
Totality Lasts for: 4:07

Veteran's Park
1005 N Detroit Ave, Russellville
Totality Starts at: 1:50pm CDT
Totality Lasts for: 4:10

Morrilton, AR
Totality Starts at: 1:50pm CDT
Totality Lasts for: 4:12
Totality Ends at: 1:54pm CDT
NWS Office: www.weather.gov/lzk
City Website: cityofmorrilton.com

Cedar Falls Overlook
Petit Jean Mountain Rd
Morrilton, AR 72110
cedar-falls-overlook.edan.io

Cherokee Park
1 Quincy Dr, Morrilton, AR 72110
Totality Starts at: 1:50pm CDT
Totality Lasts for: 4:12

Depot Museum
101 E Railroad Ave
Morrilton, AR 72110
cityofmorrilton.com

Holla Bend National Wildlife Refuge
Pottsville, AR 72858
Totality Starts at: 1:49pm CDT
Totality Lasts for: 4:16

Morrilton City Park
100-, 298 City Park Dr
Morrilton, AR 72110
cityofmorrilton.com

Never gaze at the sun without eye protection. Only remove eye protection during 100% totality.

Museum of Automobiles
8 Jones Ln
Morrilton, AR 72110
www.museumofautos.com

Petit Jean Overlook and Gravesite
Stouts Point
Morrilton, AR 72110
www.arkansasstateparks.com

Petit Jean State Park
1285 Petit Jean Mountain Rd, Morrilton, AR
arkansasstateparks.com/parks/petit-jean-state-park
Totality Lasts for: 4:15

Seven Hollows Trailhead
Petit Jean Mountain Rd
Morrilton, AR 72110
www.arkansasstateparks.com

Conway, AR
Totality Starts at: 1:51pm CDT
Totality Lasts for: 3:55
Totality Ends at: 1:55pm CDT
NWS Office: www.weather.gov/lzk
City Website: conwayarkansas.gov

Beaverfork Park Lake
20 Kinley Dr
Conway, AR 72032
conwayarkansas.gov/parks/

Cadron Settlement Park
6200 Hwy 319 W
Conway, AR 72034
conwayarkansas.gov/parks/

Camp Robinson State Wildlife Management Area
331 Clinton Road, Conway, AR 72032
Totality Starts at: 1:51pm CDT
Totality Lasts for: 3:32

Curtis Walker Park
Conway, AR 72032
conwayarkansas.gov/parks/
Totality Lasts for: 3:54

Faulkner County Museum
801 Locust St
Conway, AR 72034
www.faulknercounty.org

Gatling Park
2325 Tyler St
Conway, AR 72034
conwayarkansas.gov/parks/

Holland Bottoms State Wildlife Management Area
Cabot, AR 72023
Totality Starts at: 1:52pm CDT
Totality Lasts for: 2:26

Toad Suck Park
93 Park Rd
Bigelow, AR 72016
recreation.gov/camping/campgrounds/232721

Woolly Hollow State Park
82 Woolly Hollow Rd, Greenbrier, AR 72058
arkansasstateparks.com
Totality Lasts for: 4:01

Clinton, AR
Totality Starts at: 1:51pm CDT
Totality Lasts for: 4:15
Totality Ends at: 1:55pm CDT
NWS Office: www.weather.gov/lzk
City Website: clintonark.com

Archey Fork Park
Archey Fork Rd, Clinton, AR 72031
Totality Starts at: 1:51pm CDT
Totality Lasts for: 4:15

Cherokee Campground Recreation Area
Higden, AR 72067
recreation.gov/camping/campgrounds/10054678
Totality Lasts for: 4:08

Choctaw Park
3848 Ar 330 Hwy E
Clinton, AR 72031
recreation.gov/camping/campgrounds/232549

Dam Site Recreation Area - Greers Ferry
315 Heber Springs Rd N
Heber Springs, AR 72543
www.arkansas.com/heber-springs

Devils Fork Recreation Area
73 Devil's Fork Rd
Greers Ferry, AR 72067
recreation.gov/camping/campgrounds/232573

Indian Rock Cave & Trail
337 Snead Dr
Fairfield Bay, AR 72088
www.arkansas.com/fairfield-bay

Sandy Beach
W Front St, Heber Springs, AR 72543
Totality Starts at: 1:52pm CDT
Totality Lasts for: 4:05

South Fork Nature Center
962, Bachelor Rd, Clinton, AR 72031
Totality Starts at: 1:51pm CDT
Totality Lasts for: 4:14

Sugar Loaf Campground
1389 Resort Rd
Higden, AR 72067
recreation.gov/camping/campgrounds/232713

Sugarloaf Mountain
Trailhead Rd
Heber Springs, AR 72543
www.sugarloafheritagecouncil.org

Van Buren Recreation Area
4350 Hwy 330 S
Fairfield Bay, AR 72088
recreation.gov/camping/gateways/537

Little Rock/NLR, AR
Totality Starts at: 1:51pm CDT
Totality Lasts for: 2:38
Totality Ends at: 1:54pm CDT
NWS Office: www.weather.gov/lzk
City Website: www.littlerock.gov

AGFC Witt Stephens Jr. Central Arkansas Nature
602 President Clinton Ave
Little Rock, AR 72201
www.agfc.com

Allsopp Park
3700 Cedar Hill Rd
Little Rock, AR 72202
www.littlerock.org

Arkansas Inland Maritime Museum
120 Riverfront Park Dr
North Little Rock, AR 72114
aimmuseum.org

Arkansas State Capitol
500 Woodlane St
Little Rock, AR 72201
www.sos.arkansas.gov

Boyle Park
2000 Boyle Park Rd
Little Rock, AR 72204
www.littlerock.org

Burns Park RV Park and Campground
4101 Arlene Laman Dr
North Little Rock, AR 72118
nlr.ar.gov

Emerald Park
3098 W Scenic Dr
North Little Rock, AR 72118
nlr.ar.gov/departments/parks-and-recreation/

Esse Purse Museum & Store
1510 Main St
Little Rock, AR 72202
essepursemuseum.com

Historic Arkansas Museum
200 E 3rd St
Little Rock, AR 72201
www.arkansasheritage.com

Julius Breckling Riverfront Park
400 President Clinton Ave
Trails & overlooks along the Arkansas River
Little Rock, AR 72201

Knoop Park
20 Ozark Point
Little Rock, AR 72202
www.littlerock.org

Lake Sylvia State Park
810 AR-324, Perryville, AR 72126
Totality Starts at: 1:50pm CDT
Totality Lasts for: 4:00

Little Rock Central High School Historic Site
2120 W Daisy L Gatson Bates Dr, Little Rock
Totality Starts at: 1:51pm CDT
Totality Lasts for: 2:30

Little Rock North / Jct. I-40 KOA Journey
7820 Kampground Way
North Little Rock, AR 72118
koa.com/campgrounds/little-rock/

Never gaze at the sun without eye protection. Only remove eye protection during 100% totality.

Little Rock Zoo
1 Zoo Dr
Little Rock, AR 72205
www.littlerockzoo.com

MacArthur Park
1000 McMath Ave
Little Rock
AR 72202

Mosaic Templars Cultural Center
501 W 9th St
Little Rock, AR 72201
www.arkansasheritage.com

Murray Park
5900 Rebsamen Park Rd
Little Rock
AR 72202

Museum of Discovery
500 President Clinton Ave
Little Rock, AR 72201
www.museumofdiscovery.org

Old State House Museum
300 W Markham St
Little Rock, AR 72201
www.arkansasheritage.com

Pinnacle Mountain State Park
11901 Pinnacle Valley Rd, Roland, AR 72135
arkansasstateparks.com/pinnaclemountain/
Totality Lasts for: 3:22

The Old Mill
3800 Lakeshore Dr
North Little Rock, AR 72116
nlr.ar.gov

William J. Clinton Library and Museum
1200 President Clinton Ave
Little Rock, AR 72201
www.clintonlibrary.gov

Mountain Home, AR
Totality Starts at: 1:53pm CDT
Totality Lasts for: 3:10
Totality Ends at: 1:56pm CDT
NWS Office: www.weather.gov/lzk
www.cityofmountainhome.com

Arkansas Grand Canyon
AR-7, Jasper, AR 72641
Totality Starts at: 1:51pm CDT
Totality Lasts for: 2:27

Buffalo National River
170 Ranger Road, St. Joe, AR 72675
Totality Starts at: 1:51pm CDT
Totality Lasts for: 3:29

Bull Shoals - White River State Park
153 Dam Overlook Ln, Lakeview, AR 72642
arkansasstateparks.com/bullshoalswhiteriver/
Totality Lasts for: 2:34

Bull Shoals Caverns
1011 C S Woods Blvd
Bull Shoals, AR 72619
bullshoalscaverns.com

Cooper Park
1101 Spring St # 4
Mountain Home, AR 72653
www.cityofmountainhome.com

Cranfield Campground
720 Cranfield Rd
Mountain Home, AR 72653
recreation.gov/camping/campgrounds/232560

Cotter Spring
321 Big Spring Pkwy, Cotter, AR 72626
Totality Starts at: 1:52pm CDT
Totality Lasts for: 3:04

Dam-Quarry Campground - Norfork Lake
170 Picnic Dr
Mountain Home, AR 72653
recreation.gov/camping/campgrounds/232565

Gamaliel Campground
1860 Fout Rd
Gamaliel, AR 72537
recreation.gov/camping/campgrounds/232591

Hickory Park
466 South Hickory St
Mountain Home, AR 72653
www.cityofmountainhome.com

Lakeview Campground
450 Boat Dock Rd
Lakeview, AR 72642
recreation.gov/camping/campgrounds/232619

Norfork National Fish Hatchery
1414 State Hwy 177
Mountain Home, AR 72653
www.fws.gov/fish-hatchery/norfork

SonLight Campground and Cabins
76 Marion County 8119, Flippin, AR 72634
www.sonlightcampgroundcabins.com
Totality Lasts for: 2:30

Mountain View, AR
Totality Starts at: 1:52pm CDT
Totality Lasts for: 4:14
Totality Ends at: 1:56pm CDT
NWS Office: www.weather.gov/lzk
City Website: www.cityofmtnview.org

Blanchard Springs Recreation Area
Mountain View
AR 72533
recreation.gov/camping/campgrounds/250010

Blanchard Springs Caverns
704 Blanchard Springs Road
Fifty-Six, AR 72533
fs.usda.gov/recmain/osfnf/recreation

City Park Stone Ampitheater
Mountain View
AR 72560
www.cityofmtnview.org

Mirror Lake Waterfall
Fifty-Six, AR 72533
www.arkansas.com/norfork/
Totality Lasts for: 4:08

Ozark Folk Center State Park
1032 Park Ave, Mountain View, AR 72560
Totality Starts at: 1:52pm CDT
Totality Lasts for: 4:13

Swinging Bridge
476-448 Swinging Bridge Rd
Mountain View
AR 72560

Third Lookout AR Hwy 9
Mountain View, AR 72560
Totality Starts at: 1:52pm CDT
Totality Lasts for: 4:11

Washington Street Park
314 King St
Mountain View, AR 72560
www.cityofmtnview.org

Batesville, AR
Totality Starts at: 1:53pm CDT
Totality Lasts for: 4:03
Totality Ends at: 1:57pm CDT
NWS Office: www.weather.gov/lzk
www.cityofbatesville.com

Batesville City Park
Batesville
AR 72501
www.batesvilleparks.com

Jacksonport State Park
Newport, AR 72112
www.arkansasstateparks.com/jacksonport/
Totality Lasts for: 3:16

Maxfield Park
Batesville
AR 72501
www.batesvilleparks.com

Old Independence Regional Museum
380 S 9th St
Batesville, AR 72501
www.oirm.org

Never gaze at the sun without eye protection. Only remove eye protection during 100% totality.

Ash Flat, AR
Totality Starts at: 1:53pm CDT
Totality Lasts for: 4:12
Totality Ends at: 1:57pm CDT
NWS Office: www.weather.gov/lzk
City Website: www.ash-flat.com

Davidsonville Historic State Park
8047 Hwy 166 S, Pocahontas, AR 72455
Totality Starts at: 1:54pm CDT
Totality Lasts for: 4:01

Grand Gulf State Park
State Hwy W, Koshkonong, MO 65692
mostateparks.com/park/grand-gulf-state-park
Totality Lasts for: 3:48

Jim Hinkle Spring River State Fish Hatchery
895 State Hwy 342, Mammoth Spring, AR 72554
Totality Starts at: 1:54pm CDT
Totality Lasts for: 4:03

Lake Charles State Park
3705 AR-25, Powhatan, AR 72458
arkansasstateparks.com/lakecharles/
Totality Lasts for: 4:00

Mammoth Spring State Park
17 US-63, Mammoth Spring, AR 72554
arkansasstateparks.com/mammothspring/
Totality Lasts for: 4:00

Spring River Oak Campground
1868 River Oaks Trail
Mammoth Spring, AR 72554
www.springriveroaks.us

Missouri

Poplar Bluff, MO
Totality Starts at: 1:56pm CDT
Totality Lasts for: 4:08
Totality Ends at: 2:00pm CDT
NWS Office: www.weather.gov/pah
City Website: www.poplarbluff-mo.gov

Beaver Springs Campground
Highway HH, Piedmont, MO 63957
www.beaverspringscamp.com
Totality Lasts for: 3:48

Camelot RV Campground/RV Park
100 Camelot Dr
Poplar Bluff, MO 63901
www.camelotrvcampground.com

Holliday Landing Campground
4309 Wayne Route F
Greenville, MO 63944
www.hollidaylanding.com

Lake Wappapello State Park
MO-172, Williamsville, MO 63967
mostateparks.com/park/lake-wappapello-state-park
Totality Lasts for: 4:11

McLane Park
474 Missouri W
Poplar Bluff, MO 63901
www.poplarbluff-mo.gov

Mingo National Wildlife Refuge
24279 MO-51
Puxico, MO 63960
www.fws.gov/refuge/mingo

Mo-Ark Regional Railroad Museum
303 S Moran St
Poplar Bluff, MO 63901
www.poplarbluff-mo.gov/205/Railroad-Museum

Ozark National Scenic Riverways Park
404 Watercress Rd, Van Buren, MO 63965
www.nps.gov/ozar/index.htm
Totality Lasts for: 3:43

Piedmont Park - Clearwater Lake
821 Co Rd 418, Piedmont, MO 63957
recreation.gov/camping/campgrounds/232670
Totality Lasts for: 3:45

Poplar Bluff Museum
1010 N Main St
Poplar Bluff, MO 63901
www.pbmuseum.org

Ray Clinton Park
200, 272 Park Ave
Poplar Bluff, MO 63901
www.poplarbluff-mo.gov

Redman Creek Recreation Area
Wappapello, MO 63966
recreation.gov/camping/campgrounds/233690
Totality Lasts for: 4:10

Sam A. Baker State Park
5580 MO-143, Patterson, MO 63956
mostateparks.com/park/sam-baker-state-park
Totality Lasts for: 3:51

Sportsmans Park
1301 Black River Ind Park Rd
Poplar Bluff, MO 63901
www.poplarbluff-mo.gov

Wolf Creek Bike Trail
Co Rd 429
Poplar Bluff, MO 63901

Dexter, MO
Totality Starts at: 1:57pm CDT
Totality Lasts for: 3:53
Totality Ends at: 2:01pm CDT
NWS Office: www.weather.gov/pah
City Website: visitdexter.com

Boon Park
Dexter, MO 63841
www.dexterpark-rec.com
Totality Lasts for: 3:53

Bloomfield Park
905 S Prairie St, Bloomfield, MO 63825
Totality Starts at: 1:57pm CDT
Totality Lasts for: 3:57

Chalk Bluff Battlefield Park
County Road 368 St, Piggott, AR 72454
Totality Starts at: 1:56pm CDT
Totality Lasts for: 3:29

Dexter Welcome Center Depot and Museum
10 W South Main St
Dexter, MO 63841
visitdexter.com

Morris State Park
State Hwy WW, Campbell, MO 63933
mostateparks.com/park/morris-state-park
Totality Lasts for: 3:24

The National Stars & Stripes Museum/Library
17377 Stars and Stripes Way, Bloomfield, MO
www.nssml.org
Totality Lasts for: 3:57

Sikeston, MO
Totality Starts at: 1:58pm CDT
Totality Lasts for: 3:31
Totality Ends at: 2:01pm CDT
NWS Office: www.weather.gov/pah
City Website: www.sikeston.org

Fort Defiance State Park
US-60 &, US-62, Cairo, IL 62914
nps.gov/places/fort-defiance-state-park.htm
Totality Lasts for: 2:57

General Watkins State Park
Field Rd, Benton, MO 63736
Totality Starts at: 1:58pm CDT
Totality Lasts for: 3:57

Hunter-Dawson State Historic Site
312 Dawson Rd, New Madrid, MO 63869
Totality Starts at: 1:58pm CDT
Totality Lasts for: 2:20

New Madrid Historical Museum
1 Main St, New Madrid, MO 63869
Totality Starts at: 1:58pm CDT
Totality Lasts for: 2:17

Sikeston Depot Museum & Cultural Center
116 W Malone Ave
Sikeston, MO 63801
sikestondepot.com

Never gaze at the sun without eye protection. Only remove eye protection during 100% totality.

Towosahgy State Historic Site
Co Rd 502, East Prairie, MO 63845
Totality Starts at: 1:59pm CDT
Totality Lasts for: 1:39

Tywappity Conservation Land
County RD220,, Chaffee, MO 63740
mdc.mo.gov/discover-nature
Totality Lasts for: 4:04

Cape Girardeau, MO
Totality Starts at: 1:58pm CDT
Totality Lasts for: 4:06
Totality Ends at: 2:02pm CDT
NWS Office: www.weather.gov/pah
www.cityofcapegirardeau.org

Battle of Pilot Knob State Historic Site
118 Maple St, Pilot Knob, MO 63663
Totality Starts at: 1:57pm CDT
Totality Lasts for: 2:40

Bollinger Mill State Historic Site
113 Bollinger Mill Rd, Burfordville
Totality Starts at: 1:57pm CDT
Totality Lasts for: 4:10

Capaha Park
1400 Broadway St
Cape Girardeau, MO 63701
www.cityofcapegirardeau.org

Cape Girardeau County Park
2400 County Park Dr
Cape Girardeau, MO 63701
www.cityofcapegirardeau.org

Cape Rock Park
E Cape Rock Dr
Cape Girardeau, MO 63701
www.cityofcapegirardeau.org

Crisp Museum
518 S Fountain St
Cape Girardeau, MO 63703
semo.edu/museum/

Elephant Rocks State Park
7390, 7406 MO-21, Belleview, MO 63623
mostateparks.com/park/elephant-rocks-state-park
Totality Lasts for: 2:20

Historic Fort D
920 Fort St
Cape Girardeau, MO 63701
www.cityofcapegirardeau.org

Jackson City Park
Jackson, MO 63755
Totality Starts at: 1:58pm CDT
Totality Lasts for: 4:09

Johnson's Shut-Ins State Park Campground
Middle Brook, MO 63656
Totality Starts at: 1:57pm CDT
Totality Lasts for: 2:17

Kiwanis Park
2100 Rotary Dr
Cape Girardeau, MO 63701
www.cityofcapegirardeau.org

Taum Sauk Lookout Tower
Lookout Tower Rd, Ironton, MO 63650
Totality Starts at: 1:57pm CDT
Totality Lasts for: 2:40

The Glenn House
325 S Spanish St
Cape Girardeau, MO 63703
www.glennhouse.org/

Tower Rock Natural Area
Pcr 460, Frohna, MO 63748
Totality Starts at: 1:58pm CDT
Totality Lasts for: 4:06

Trail of Tears State Park Scenic Overlook
429 Moccasin Springs Rd, Jackson, MO 63755
mostateparks.com/park/trail-tears-state-park
Totality Lasts for: 4:09

World's Largest Fountain Drink Cup
425 S Mt Auburn Rd
Cape Girardeau, MO 63703
Totality Lasts for: 4:05

Kentucky

Paducah, KY
Totality Starts at: 2:00pm CDT
Totality Lasts for: 2:00 Northwest
Totality Ends at: 2:00pm CDT
NWS Office: www.weather.gov/pah
City Website: www.paducahky.gov

Bob Noble Park
2801 Park Ave
Paducah, KY 42001
www.paducahky.gov/noble-park

Carson Park
300 N 30th St
Paducah, KY 42001
www.paducahky.gov

Columbus Belmont State Park
350 Park Road, Columbus, KY 42032
parks.ky.gov
Totality Lasts for: 1:37

Fort Jefferson Hill Park and Memorial Cross
Wickliffe, KY 42087
Totality Starts at: 1:59pm CDT
Totality Lasts for: 2:44

Fort Massac State Park
1308 E 5th St, Metropolis, IL 62960
dnr.illinois.gov
Totality Lasts for: 2:29

Historic Riverfront
27 The Foot of Broadway
Paducah, KY 42001
www.paducahky.gov

Lloyd Tilghman House & Civil War Museum
631 Kentucky Ave
Paducah, KY 42001
www.paducahky.gov

Paducah Railroad Museum
Washington St & Marine Way
Paducah, KY 42003
www.paducahrr.org/

Paducah Wall to Wall
200-298 S Water St
Paducah, KY 42003
paducahwalltowall.com

The National Quilt Museum
215 Jefferson St
Paducah, KY 42001
www.quiltmuseum.org

Superman / Lois Lane Statue
517 Market St
Metropolis, IL 62960
supermuseum.com

Wickliffe Mounds State Historic Site
94 Green St, Wickliffe, KY 42087
Totality Starts at: 1:59pm CDT
Totality Lasts for: 2:47

Henderson, KY
Totality Starts at: 2:02pm CDT
Totality Lasts for: 2:32
Totality Ends at: 2:05pm CDT
NWS Office: www.weather.gov/pah
City Website: www.hendersonky.gov

Atkinson Park
1801 N Elm St
Henderson, KY 42420
www.hendersonky.gov

John James Audubon State Park
3100 US Hwy 41, Henderson, KY 42420
parks.ky.gov
Totality Lasts for: 2:41

Sandy Lee Watkins Park
16040 KY-351, Henderson, KY 42420
Totality Starts at: 2:03pm CDT
Totality Lasts for: 1:00

Never gaze at the sun without eye protection. Only remove eye protection during 100% totality.

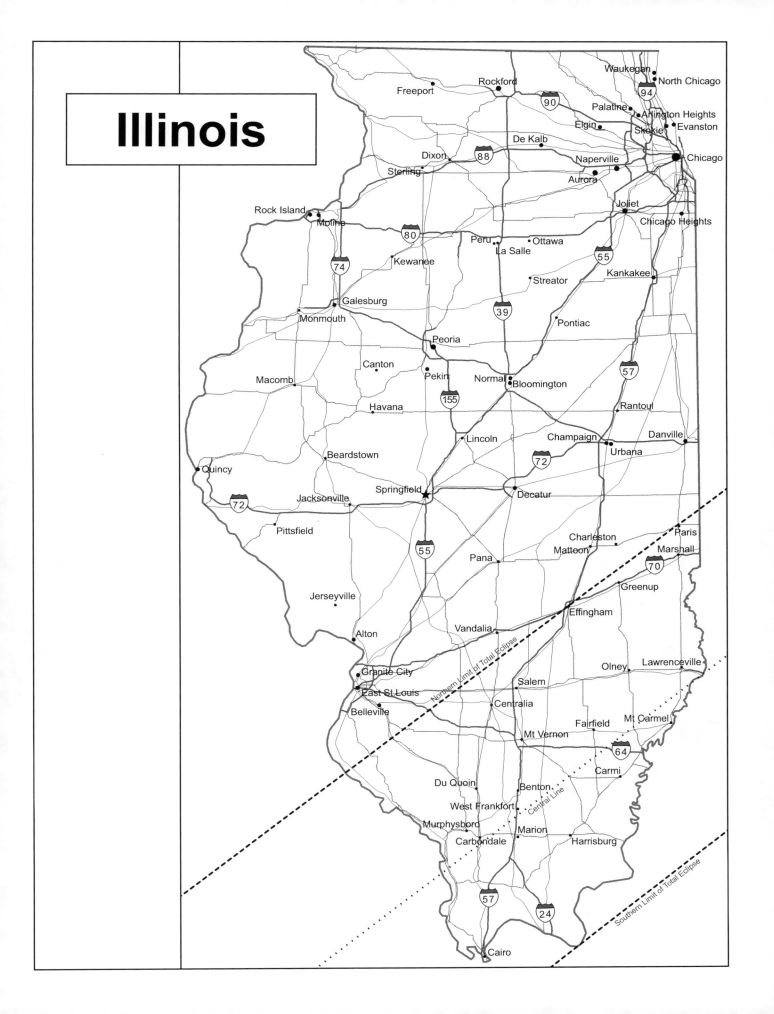

Illinois

Featured Places to Eat

Agave Mexican Restaurant in Mt. Vernon is focused on Mexican entrees made from fresh ingredients. Customers can choose from an extensive selection of dishes that include enchiladas, steaks, fajitas, quesadillas, burritos, pork dishes, and seafood specialties. Vegetarian options are also available. Meals are usually served with rice and black beans. The total eclipse of the sun will last nearly 4 minutes in this area, be sure to look up and enjoy it! Located at 300 S 44th St, (618) 244-7454

Harbaugh's Cafe in Carbondale has an extensive and appetizing menu that includes a variety of breakfast and lunch options. They have specialty egg dishes such as Egg Plus, Biscuits & Gravy, Corned Beef Hash, and Eggs Benedict. For those who enjoy omelettes, the cafe serves a range of three-egg omelettes with different fillings. They also have pancakes, French toast, soups, salads and sandwiches.
For over 4 minutes, day turns to night in Carbondale. Get your meal and be ready to marvel! Located at 901 S Illinois Ave B, (618) 351-9897

Jack Russell Fish Co. in Benton is a seafood lover's paradise. Among their favorites are Red Beans and Rice Dinner, and Jack's Big Cod, a generous 13 oz. portion. The "Hot Wings" featuring fresh-cut jumbo chicken wings are a must-try. In Benton, a 4 minute solar eclipse performance awaits. Order your food early and get ready for the awe-inspiring display. Located at 106 E Main St, (618) 439-3474

Mackie's Pizza in Harrisburg sells a mouthwatering selection of pizzas with options for thin crust, deep dish, and cauliflower crust. Customers can indulge in specialty pizzas such as the "Harrisburg Special," loaded with pepperoni, sausage, hamburger, bacon, onions, peppers, olives, mushrooms, apples, and anchovies. In addition to pizzas, Mackie's serves a variety of toasted sandwiches, including the "Super Sub" and "Stromboli." Nearly 4 minutes of totality will blaze in the sky here on eclipse day. Located at 502 E Poplar St, (618) 252-6368

Quatro's in Carbondale offers a wide range of delectable choices, including their famous deep pan and thin-style pizzas with customizable toppings. Specialty pizzas like Quatro's Challenge and Philly Cheese Steak Pizza are highlights of the menu. Alongside pizzas, Quatro's serves savory pasta dishes such as Lasagna and Chicken Alfredo, as well as signature sandwiches like the Sooper Reuben and Frisco Steak Melt. For the 2024 eclipse, Carbondale is near the centerline once again and visitors here will enjoy over 4 minutes of totality. Located at 218 W Freeman St, (618) 549-5326

Whiffle Boy's Pizza in Murphysboro has a delicious variety of hand-tossed or thin crust pizzas. Specialty pizzas include options like "The Breakfast Pizza," featuring scrambled eggs, sausage, and bacon, and "BBQ Chicken" with red onion and BBQ sauce. Whiffle Boy's Pizza also has oven-toasted subs, fresh salads, and desserts. The moon and sun will perform a total solar eclipse for over 4 minutes in this area. Located at 2039 Walnut St, (618) 687-9433

Eclipse Track Notes - IL

Cities near the northern and southern limits
East St. Louis, Illinois is outside the limits of the total eclipse. From St. Louis head southeast on Interstate I-64 toward Mt. Vernon or go east on US-50 to Salem or Olney. Carbondale and West Frankfort are close to the center of the eclipse and easily accessible by continuing from I-64 to I-57 south. State Highway IL-13 crosses I-57 south of West Frankfort and connects Carbondale and Marion. Effingham is right on the northern limit of the total eclipse. Head east on I-70 toward Terre Haute, Indiana, or south on I-57 toward Salem, Illinois, from Effingham for a longer duration of totality.

Mobility
In the north, I-70 will take you into the total eclipse just east of Effingham. I-57 runs from Effingham to the southern tip of Illinois. I-64 from St. Louis crosses I-57 at Mt. Vernon and continues east to just north of Evansville, Indiana. US-50 due east connects St. Louis to Salem and Olney, Illinois, and Vincennes, Indiana.

Murphysboro, IL
Totality Starts at: 1:59pm CDT
Totality Lasts for: 4:04
Totality Ends at: 2:03pm CDT
NWS Office: www.weather.gov/pah
City Website: www.murphysboro.com

17th Street Barbecue
32 N 17th St, (618) 684-3722
BBQ, Murphysboro
Eclipse Day Monday, 11am - 9pm

Iam Java Coffee House
715 N 14th St, (618) 534-5341
Coffee Shop, Murphysboro
Eclipse Day Monday, 6am - 2pm

Martel's Pizza
706 Walnut St, (618) 684-1111
Pizza, Murphysboro
Eclipse Day Monday, 11am - 9pm

Whiffle Boy's Pizza
2039 Walnut St, (618) 687-9433
Pizza, Murphysboro
Eclipse Day Monday, 11am - 9pm

Carbondale, IL
Totality Starts at: 1:59pm CDT
Totality Lasts for: 4:09
Totality Ends at: 2:03pm CDT
NWS Office: www.weather.gov/pah
Eclipse Website: eclipse.siu.edu

Bandana's Bar-B-Q Carbondale II
309 E Main St, (618) 490-1777
BBQ, Carbondale
Eclipse Day Monday, 11am - 9pm

Chango's Bar & Grill
519 S Illinois Ave, Carbondale, IL 62901
Mexican, Carbondale
Eclipse Day Monday, 11am - 11pm

Dale's Burger Shack
1709 W Main St, (618) 521-5723
American, Carbondale
Eclipse Day Monday, 9am - 5pm

Don Sol Mexican Grill
715 N Giant City Rd, (618) 351-0002
Mexican, Carbondale
Eclipse Day Monday, 11am - 9:30pm

Don Taco
780 E Grand Ave, (618) 549-3777
Mexican, Carbondale
Eclipse Day Monday, 11am - 2am

El Paisano Mexican Grill
1925 W Main St, (618) 319-1685
Mexican, Carbondale
Eclipse Day Monday, 11am - 9pm

Fujiyama Japanese Cuisine
225 N Giant City Rd, (618) 549-2000
Japanese, Carbondale
Eclipse Day Monday, 11am - 2:30pm, 4:30–9:30

Harbaugh's Cafe
901 S Illinois Ave B, (618) 351-9897
American, Carbondale
Eclipse Day Monday, 7am - 2pm

Hunan
710 E Main St, (618) 529-1108
Chinese, Carbondale
Eclipse Day Monday, 11am - 8pm

Pagliai's Pizza
509 S Illinois Ave, (618) 657-7009
Pizza, Carbondale
Eclipse Day Monday, 11am - 9pm

Primo's Pizza
604 E Park St, (618) 351-9999
Pizza, Carbondale
Eclipse Day Monday, 11am - 11pm

Qin Guan Restaurant
1285 E Main St, (618) 351-1222
Chinese Buffet, Carbondale
Eclipse Day Monday, 11am - 9pm

Quatro's Deep Pan Pizza
218 W Freeman St, (618) 549-5326
Pizza, Carbondale
Eclipse Day Monday, 11am - 10pm

Sergio's Mexican Restaurant
1160 E Main St, (618) 529-1173
Mexican, Carbondale
Eclipse Day Monday, 10am - 10pm

Never gaze at the sun without eye protection. Only remove eye protection during 100% totality.

Tequila's
100 N Glenview Dr #2275, (618) 457-4026
Mexican, Carbondale
Eclipse Day Monday, 11am - 10pm

Yamato Steak House Of Japan
1013 E Main St, (618) 351-6888
Japanese, Carbondale
Eclipse Day Monday, 11am - 9:30pm

Marion, IL
Totality Starts at: 1:59pm CDT
Totality Lasts for: 4:07
Totality Ends at: 2:03pm CDT
NWS Office: www.weather.gov/pah
City Website: cityofmarionil.gov

Baan Thai
2406 Williamson County Pkwy, (618) 998-1555
Thai, Marion
Eclipse Day Monday, 11am - 2:45pm, 4–8:45pm

Bennie's Italian Foods
309 N Market St, (618) 997-6736
Italian, Marion
Eclipse Day Monday, 10:30am - 8pm

Chop Kitchen
2703 17th St, (618) 422-2100
Sushi, Marion
Eclipse Day Monday, 11am - 9:30pm

Don Luna Mexican Restaurant
407 E Main St, (618) 997-9399
Mexican, Marion
Eclipse Day Monday, 10:30am - 8pm

Don Sol Mexican Grill
2800 17th St, (618) 997-8181
Mexican, Marion
Eclipse Day Monday, 11am - 9:30pm

El Ranchito Restaurant
809 N Court St, (618) 969-7149
Mexican, Marion
Eclipse Day Monday, 11am - 9:30pm

La Fiesta Mexican Restaurant
1000 N Carbon St, (618) 993-0028
Mexican, Marion
Eclipse Day Monday, 10:30am - 10pm

Let's Beef Frank Food Truck
901-1099 N Russell St, (618) 922-3482
American, Marion
Eclipse Day Monday, 10:30am - 4pm

Tequilas Mexican Restaurant
1906 W Coolidge Ave, (618) 997-0162
Mexican, Marion
Eclipse Day Monday, 10am - 9pm

Thai-D Classic Thai Cuisine
2801 Civic Cir Blvd Suite.6, (618) 997-6470
Thai, Marion
Eclipse Day Monday, 11am - 2:30pm, 4–8pm

Triple E Bar-B-Q II
706 Robinson Dr, (618) 997-5559
BBQ, Marion
Eclipse Day Monday, 10am - 8pm

Harrisburg, IL
Totality Starts at: 2:00pm CDT
Totality Lasts for: 3:54
Totality Ends at: 2:04pm CDT
NWS Office: www.weather.gov/pah
City: www.thecityofharrisburgil.com

Bar-B-Q Barn
632 N Main St, (618) 252-6190
BBQ, Harrisburg
Eclipse Day Monday, 7am - 7pm

El Ranchito Restaurant
303 S Commercial St #3, (618) 252-4203
Mexican, Harrisburg
Eclipse Day Monday, 11am - 9:30pm

Johnson's Southern Style Bar-B-Q
700 E Walnut St, (618) 252-0477
BBQ, Harrisburg
Eclipse Day Monday, 10am - 7pm

KEITHS Cafe Bar and Grill
601 N Commercial St, (618) 252-9040
American, Harrisburg
Eclipse Day Monday, 10am - 12am

Mackie's Pizza
502 E Poplar St, (618) 252-6368
Pizza, Harrisburg
Eclipse Day Monday, 11am - 8:30pm

Morello's Restaurant & Catering
217 E Poplar St, (618) 252-2300
Italian, Harrisburg
Eclipse Day Monday, 11am - 9pm

Peking Palace Restaurant
303 S Commercial St #11, (618) 252-1401
Chinese, Harrisburg
Eclipse Day Monday, 11am - 9pm

steam cafe
10 S Cherry St #3, (618) 926-5477
Coffee Shop, Harrisburg
Eclipse Day Monday, 7am - 4pm

Tequilas Mexican Restaurant- Harrisburg
507 N Commercial St, (618) 252-4267
Mexican, Harrisburg
Eclipse Day Monday, 11am - 10pm

West Frankfort, IL
Totality Starts at: 1:59pm CDT
Totality Lasts for: 4:08
Totality Ends at: 2:04pm CDT
NWS Office: www.weather.gov/pah
City: www.westfrankfort-il.com

Bonnie Cafe
15 W Frankfort Plaza, (618) 932-9777
American, West Frankfort
Eclipse Day Monday, 6am - 8pm

La Fiesta Mexican Restaurant
1402 W Main St, (618) 937-2838
Mexican, West Frankfort
Eclipse Day Monday, 10:30am - 9pm

Mike's Drive-In
1007 W Main St, (618) 932-2564
American, West Frankfort
Eclipse Day Monday, 11am - 8pm

New China Star
705 W Main St, (618) 932-3281
Chinese, West Frankfort
Eclipse Day Monday, 11am - 9pm

Pup's 212 Bar & Grill
212 E Main St, (618) 932-2225
American, West Frankfort
Eclipse Day Monday, 9am - 12am

Du Quoin, IL
Totality Starts at: 1:59pm CDT
Totality Lasts for: 3:51
Totality Ends at: 2:03pm CDT
NWS Office: www.weather.gov/pah
City: www.duquointourism.org

Corner Pocket
402 S Washington St, (618) 542-3276
American, Du Quoin
Eclipse Day Monday, 10am - 1:30am

Don Tequilas Mexican Restaurant
222 Southtowne Shopping Ctr, (618) 542-8008
Mexican, Du Quoin
Eclipse Day Monday, 11am - 9pm

Kalin's Cafe
9 E Main St, (618) 542-2228
Breakfast, Du Quoin
Eclipse Day Monday, 5am - 9pm

St. Nicholas Brewing Company
12 S Oak St, (618) 790-9212
American, Du Quoin
Eclipse Day Monday, 11am - 8pm

Benton, IL
Totality Starts at: 2:00pm CDT
Totality Lasts for: 4:05
Totality Ends at: 2:04pm CDT
NWS Office: www.weather.gov/pah
City Website: bentonil.com

Full Moon House
17 Rend Lake Rd, (618) 435-4333
Chinese, Benton
Eclipse Day Monday, 11am - 9pm

Jack Russell Fish Company
106 E Main St, (618) 439-3474
Seafood, Benton
Eclipse Day Monday, 11am - 2pm, 4–8pm

Keyaki Sushi Hibachi Steak House
100 N Central St, (618) 435-6868
Japanese, Benton
Eclipse Day Monday, 11am - 9pm

Never gaze at the sun without eye protection. Only remove eye protection during 100% totality.

La Fiesta Mexican Restaurant
110 Jackson St, (618) 435-2888
Mexican, Benton
Eclipse Day Monday, 10:30am - 9pm

Mr D's
219 N Main St, (618) 439-9647
American, Benton
Eclipse Day Monday, 8am - 10pm

Pop's BBQ
322 N Main St, (618) 438-7000
BBQ, Benton
Eclipse Day Monday, 11am - 7pm

SI Pho
601 W Main St, (618) 663-3622
Vietnamese, Benton
Eclipse Day Monday, 11am - 2:30pm, 4:30–9pm

Tequila's Mexican Restaurant
721 W Washington St, (618) 435-3112
Mexican, Benton
Eclipse Day Monday, 11am - 9pm

The Cozy Table
1514 N Main St, (618) 435-3081
American, Benton
Eclipse Day Monday, 6am - 3pm

Mt Vernon, IL
Totality Starts at: 2:00pm CDT
Totality Lasts for: 3:41
Totality Ends at: 2:04pm CDT
NWS Office: www.weather.gov/pah
City Website: www.mtvernon.com

Agave Mexican Restaurant
300 S 44th St, (618) 244-7454
Mexican, Mt Vernon
Eclipse Day Monday, 11am - 9:30pm

Bonnie Cafe
100 Aviation Dr, (618) 242-4070
Breakfast, Mt Vernon
Eclipse Day Monday, 6am - 8pm

Double Overtime Grill
222 Potomac Blvd, (618) 241-6959
American, Mt Vernon
Eclipse Day Monday, 11:30am - 10pm

El Rancherito of Mt Vernon, IL
4303 Broadway St, (618) 244-6121
Mexican, Mt Vernon
Eclipse Day Monday, 11am - 9pm

Fujiyama Japanese Cuisine
4809 Broadway St, (618) 315-6608
Japanese, Mt Vernon
Eclipse Day Monday, 11am - 2:30pm, 4:30–9:30

La Fiesta Mexican Restaurant
901 Prairie Ave, (618) 315-6688
Mexican, Mt Vernon
Eclipse Day Monday, 11am - 9pm

La Fuente Mexican Restaurant
222 S 9th St, (618) 315-6369
Mexican, Mt Vernon
Eclipse Day Monday, 11am - 9pm

Show Down Steakhouse & Saloon
333 Potomac Blvd, (618) 316-7006
Steak, Mt Vernon
Eclipse Day Monday, 11am - 9pm

The Farmhouse Bakery
1812 Broadway St, (618) 816-4001
Breakfast, Mt Vernon
Eclipse Day Monday, 7am - 5:30pm

The Frosty Mug Bar & Grill
1113 Salem Rd, (618) 242-3372
American, Mt Vernon
Eclipse Day Monday, 9am - 11pm

Trackside Bar & Grill
201 Broadway St, (618) 315-6090
American, Mt Vernon
Eclipse Day Monday, 11am - 9pm

Waffle Company
4115a Broadway St, (618) 242-6047
Breakfast, Mt Vernon
Eclipse Day Monday, 5:30am - 2:30pm

Fairfield, IL
Totality Starts at: 2:01pm CDT
Totality Lasts for: 4:01
Totality Ends at: 2:05pm CDT
NWS Office: www.weather.gov/pah
City: cityoffairfieldillinois.com

Barbwire Grill
909 W Delaware St, (618) 842-2532
American, Fairfield
Eclipse Day Monday, 6am - 8pm

Classic Pizza & Pasta
215 E Main St, (618) 847-8181
Pizza, Fairfield
Eclipse Day Monday, 11am - 9pm

El Mexicano Restaurant
5 Williamson Dr Ste 200, (618) 842-7868
Mexican, Fairfield
Eclipse Day Monday, 11am - 9pm

Five Brothers Cafe
110 Market Ave, (618) 842-3366
American, Fairfield
Eclipse Day Monday, 5:45am - 9pm

K & M Diner
410 S 1st St, (618) 516-7198
American, Fairfield
Eclipse Day Monday, 7am - 1pm

La Fuente Mexican Restaurant
1007 Commerce Drive, (618) 842-5000
Mexican, Fairfield
Eclipse Day Monday, 11am - 9pm

Panda Buffet
1005 Commerce Drive, (618) 847-7030
Chinese, Fairfield
Eclipse Day Monday, 11am - 9:30pm

Taco Tierra of Fairfield
1100 W Main St, (618) 842-3377
Mexican, Fairfield
Eclipse Day Monday, 9:30am - 9pm

Mt Carmel, IL
Totality Starts at: 2:02pm CDT
Totality Lasts for: 4:03
Totality Ends at: 2:06pm CDT
NWS Office: www.weather.gov/pah
City: www.cityofmtcarmel.com

Big Jon's Lunch Box
123 W 4th St, (618) 263-3272
American, Mt Carmel
Eclipse Day Monday, 8am - 1:30pm

Little Italy's Pizza
502 N Walnut St, (618) 262-4121
Pizza, Mt Carmel
Eclipse Day Monday, 11am - 10pm

Taco Tierra of Mt. Carmel
729 N Market St, (618) 262-8226
Mexican, Mt Carmel
Eclipse Day Monday, 9am - 10pm

Tequila's Mexican Restaurant
115 W 9th St, (618) 262-5275
Mexican, Mt Carmel
Eclipse Day Monday, 11am - 9:30pm

Olney, IL
Totality Starts at: 2:02pm CDT
Totality Lasts for: 3:46
Totality Ends at: 2:06pm CDT
NWS Office: www.weather.gov/ilx
City Website: www.ci.olney.il.us

Chilly Willy's Ice Creams & Grill
416 E Main St, (618) 320-9866
American, Olney
Eclipse Day Monday, 11am - 9pm

China Capital Super Buffet
525 N West St, (618) 392-8889
Chinese, Olney
Eclipse Day Monday, 10:30am - 9pm

El Cactus Mexican Grill
200 W Main St, (618) 395-7000
Mexican, Olney
Eclipse Day Monday, 11am - 9pm

Never gaze at the sun without eye protection. Only remove eye protection during 100% totality.

Ginger Ale's Olney
612 N West St, (618) 395-4464
American, Olney
Eclipse Day Monday, 7am - 8pm

Hog N Dog BBQ
320 S West St, (618) 320-0456
BBQ, Olney
Eclipse Day Monday, 11am - 7pm

Hovey's Diner
412 E Main St, (618) 395-4683
American, Olney
Eclipse Day Monday, 6am - 7pm

Mis Tres Potrillos Olney
211 S West St, (618) 392-6415
Mexican, Olney
Eclipse Day Monday, 10am - 9pm

Ophelia's Cup
205 S Whittle Ave, (618) 392-6287
Coffee Shop, Olney
Eclipse Day Monday, 7am - 1:30pm

Pizza Fast
307 S Whittle Ave, (618) 395-7417
Pizza, Olney
Eclipse Day Monday, 11am - 7pm

Mi Casita Mexican Restaurant
1123 State St, (618) 943-5850
Mexican, Lawrenceville
Eclipse Day Monday, 11am - 9pm

Phoenix Grill
Lawrenceville, IL 62439, (618) 707-2255
Breakfast, Lawrenceville
Eclipse Day Monday, 5am - 2pm

Syd's Place Inc
1802 11th St, (618) 943-9040
American, Lawrenceville
Eclipse Day Monday, 8am - 2:30pm

Lawrenceville, IL
Totality Starts at: 2:02pm CDT
Totality Lasts for: 4:02
Totality Ends at: 2:06pm CDT
NWS Office: www.weather.gov/ilx
City Website: lawrencevilleil.org

Hoagy House
701-799 13th St, (618) 943-3916
Sandwich, Lawrenceville
Eclipse Day Monday, 10am - 3pm

Lawrenceville Drive-Inn
2002 15th St, (618) 943-2271
American, Lawrenceville
Eclipse Day Monday, 10am - 9pm

Mad Pig Bar-B-Q
2006 15th St, (812) 881-7902
BBQ, Lawrenceville
Eclipse Day Monday, 10:30am - 2pm

Featured Parks and Attractions

Dixon Springs State Park is located in Pope County, Illinois, about 10 miles west of Golconda on Illinois Route 146. Taking its name from William Dixon, one of the area's early settlers, the park's rich history includes a vibrant small community complete with a general store, post office, blacksmith shop, and several churches. The park's seven mineral-enriched springs make it a popular health spa, drawing visitors from far and wide. Camping, fishing, and hunting are popular activities at this park. The totality phase of the eclipse lasts for over three minutes at Dixon Springs State Park.

Ferne Clyffe State Park, located in Johnson County, Illinois is easily accessible via Illinois Route 37. The park's landscape features a mix of limestone bluffs, dense woodlands, intriguing cliff caves, and a number of small seasonal waterfalls that are fed by runoff from the upper bluffs. An artificial lake, Ferne Clyffe Lake, was created in the late 1950s and offers opportunities for boating, fishing and many other outdoor activities. Totality will last four minutes from this park because its very close to the center line of the path of totality.

Fort Kaskaskia State Historic Site is a 200-acre park located near Chester, Illinois. It was a strategic location during the American Revolution. The site commemorates the now-vanished frontier town of Old Kaskaskia and its role in American history. Fort Kaskaskia was the first capital of the state of Illinois after its admission to the Union in 1818. Visitors can camp, picnic, and explore what remains of the original site and immerse themselves in the stories of the settlers, soldiers, and Native Americans who once lived there. The moon/sun battle on April 8, 2024 culminates in a total solar eclipse visible for over 3 minutes from this site.

Garden of the Gods, located in Southern Illinois, is a popular tourist destination for its stunning rock formations and forest landscapes. This park offers a variety of recreational opportunities, including hiking, camping, and bird watching. The best feature of the park is the Observation Trail, a quarter-mile loop that leads visitors past the sandstone formations and incredible views, including Camel Rock and Devil's Smokestack. Picnic areas and tent and trailer campsites make this park a perfect location for both day trips and extended stays. Nearly three and a half minutes of totality will treat visitors at this park on eclipse day.

Giant City State Park is located in Southern Illinois and covers 4,000 acres. The park is a popular destination, offering hiking, horseback riding, picnicking, and a lodge and cabins. It's also home to the Giant City Stone Fort Site, a prehistoric stone fort listed on the National Register of Historic Places. This park is very close to the eclipse central line and visitors will experience over four minutes of totality.

The Super Museum, the "worlds largest collection of Superman", is located in Metropolis, Illinois - the official hometown of the Man of Steel. Opened in 1993 by Superman enthusiast and collector, Jim Hambrick, the museum houses over 70,000 items, including toys, movie props, movie and TV promotional materials, and vintage Superman costumes. There is a statue of Lois Lane only 2 blocks away. Despite his remarkable powers, Superman will not be able to intervene in the natural spectacle of the total solar eclipse occurring on April 8, 2024, when the sky will plunge into darkness for a duration of 2 minutes and 47 seconds outside this interesting museum.

The parks and attractions section in this book are organized by cities, which are highlighted in bold. Under each city, you will find a curated list of locations presented in alphabetical order. These locations may be within the city or situated within a 30-50 mile radius, encompassing the surrounding region. This method of organization allows for easy navigation and planning, ensuring you can make the most of your solar eclipse experience.

Never gaze at the sun without eye protection. Only remove eye protection during 100% totality.

Parks and Attractions

Murphysboro, IL
Totality Starts at: 1:59pm CDT
Totality Lasts for: 4:04
Totality Ends at: 2:03pm CDT
NWS Office: www.weather.gov/pah
City Website: www.murphysboro.com

Fort Kaskaskia State Historic Site
4372 Park Rd, Ellis Grove, IL 62241
Totality Starts at: 1:59pm CDT
Totality Lasts for: 3:05

General John A Logan Museum
1613 Edith St
Murphysboro, IL 62966
www.loganmuseum.org

Glenn Schlimpert Recreation Area
1769 Water Plant Rd
Murphysboro, IL 62966
Totality Lasts for: 4:01

Johnson Creek Recreation Area
Johnson Creek Rd, Ava, IL 62907
fs.usda.gov
Totality Lasts for: 3:53

Kinkaid Lake Spillway
432 N Spillway Rd
Gorham, IL 62940
www.fs.usda.gov

Lake Murphysboro State Park
52 Cinder Hill Dr, Murphysboro, IL 62966
dnr.illinois.gov
Totality Lasts for: 4:03

Randolph County State Recreation Area
4301 S Lake Dr, Chester, IL 62233
Totality Starts at: 1:59pm CDT
Totality Lasts for: 3:15

Carbondale, IL
Totality Starts at: 1:59pm CDT
Totality Lasts for: 4:09
Totality Ends at: 2:03pm CDT
NWS Office: www.weather.gov/pah
City Website: eclipse.siu.edu

Attucks Park
800 N Wall St
Carbondale, IL 62901
www.cpkd.org

Castle Park
101 Homewood Dr #99
Carbondale, IL 62902
Totality Lasts for: 4:08

Crab Orchard NWR Campground
10000-10018 Campground Dr
Carterville, IL 62918
www.fws.gov/refuge/crab-orchard

Evergreen Park
1205 W Pleasant Hill Rd
Carbondale, IL 62903
www.cpkd.org

Giant City State Park
235 Giant City Rd, Makanda, IL 62958
dnr.illinois.gov
Totality Lasts for: 4:08

Jeremy "Boo" Rochman Memorial Park
31 Homewood Dr
Carbondale, IL 62902
atlasobscura.com

Little Grand Canyon
Pomona, IL 62975
Totality Starts at: 1:58pm CDT
Totality Lasts for: 4:06

Marberry Arboretum
1398 E Pleasant Hill Rd
Carbondale, IL 62902
arboretum.siu.edu

Pomona Natural Bridge
Natural Bridge Rd, Pomona, IL 62975
Totality Starts at: 1:59pm CDT
Totality Lasts for: 4:09

The Science Center
1237 E Main St Space 1048
Carbondale, IL 62901
sciencecentersi.com

Marion, IL
Totality Starts at: 1:59pm CDT
Totality Lasts for: 4:07
Totality Ends at: 2:03pm CDT
NWS Office: www.weather.gov/pah
City Website: cityofmarionil.gov

Crab Orchard National Wildlife Refuge
8588 IL-148, Marion, IL 62959
Nature Preserve
www.fws.gov/refuge/crab-orchard

Dixon Springs State Park
982 IL-146, Golconda, IL 62938
dnr.illinois.gov
Totality Lasts for: 3:17

Ferne Clyffe State Park
90 Goreville Rd, Goreville, IL 62939
Totality Starts at: 1:59pm CDT
Totality Lasts for: 4:00

Mandala Gardens
1704 N State St
Marion, IL 62959
www.mandalagardens.org

Pyramid Park
1421 Pharaoh's Wy
Marion, IL 62959
www.marionparks.com

Ray Fosse Park
500 E Deyoung St
Marion, IL 62959
www.marionparks.com

Shawnee Bluffs Canopy Tour
635 Robinson Hill Rd
Makanda, IL 62958
www.shawneezip.com

Harrisburg, IL
Totality Starts at: 2:00pm CDT
Totality Lasts for: 3:54
Totality Ends at: 2:04pm CDT
NWS Office: www.weather.gov/pah
www.thecityofharrisburgil.com

Bell Smith Springs Scenic Area
Bell Smith Springs Rd, Ozark, IL 62972
Totality Starts at: 2:00pm CDT
Totality Lasts for: 3:38

Cave-In-Rock State Park
1 New State Park Rd, Cave-In-Rock, IL 62919
dnr.illinois.gov/parks/park.caveinrock.html
Totality Lasts for: 2:28

Garden of the Gods
Herod, IL 62947
www.fs.usda.gov
Totality Lasts for: 3:27

Illinois Iron Furnace National Historic Site
Route 146 &, 34, Elizabethtown, IL 62931
fs.usda.gov
Totality Lasts for: 2:59

Rim Rock National Recreation Trail
Eagle Creek, RR 1 Box 198B
Findlay, IL 62534
fs.usda.gov

Saline County Area Museum
1600 S Feazel St
Harrisburg, IL 62946
saline-village.edan.io

Saline County State Fish and Wildlife Area
85 Glen Jones Rd, Equality, IL 62934
Totality Starts at: 2:00pm CDT
Totality Lasts for: 3:39

Never gaze at the sun without eye protection. Only remove eye protection during 100% totality.

West Frankfort, IL
Totality Starts at: 1:59pm CDT
Totality Lasts for: 4:08
Totality Ends at: 2:04pm CDT
NWS Office: www.weather.gov/pah
www.westfrankfort-il.com

Arrowhead Lake Campground & Recreational
1600 Peterson Ave
Johnston City, IL 62951
arrowheadlakecampground.com

Frankfort Community Park
1100 E Cleveland St
West Frankfort, IL 62896
frankfortpark.com

Herrin Memorial Park
1010 N 5th St
Herrin, IL 62948
Totality Lasts for: 4:08

Du Quoin, IL
Totality Starts at: 1:59pm CDT
Totality Lasts for: 3:51
Totality Ends at: 2:03pm CDT
NWS Office: www.weather.gov/pah
www.duquointourism.org

DuQuoin State Fairgrounds
655 Executive Dr
Du Quoin, IL 62832
www.duquoinstatefair.net

Pyramid State Recreation Area
1562 Pyramid Park Rd
Pinckneyville, IL 62274
Totality Lasts for: 3:40

Washington County State Recreation Area
18500 Conservation Dr, Nashville, IL 62263
Totality Starts at: 2:00pm CDT
Totality Lasts for: 3:11

Benton, IL
Totality Starts at: 2:00pm CDT
Totality Lasts for: 4:05
Totality Ends at: 2:04pm CDT
NWS Office: www.weather.gov/pah
City Website: bentonil.com

Benton City Park
802 1st St
Benton, IL 62812
www.bentonil.com

South Sandusky Camp Ground
Benton, IL 62812
recreation.gov/camping/campgrounds/233613
Totality Lasts for: 3:59

Wayne Fitzgerrell State Recreation Area
11094 Ranger Rd, Whittington, IL 62897
Totality Starts at: 2:00pm CDT
Totality Lasts for: 3:59

Mt Vernon, IL
Totality Starts at: 2:00pm CDT
Totality Lasts for: 3:41
Totality Ends at: 2:04pm CDT
NWS Office: www.weather.gov/pah
City Website: www.mtvernon.com

Cedarhurst Center For the Arts
2600 E Richview Rd
Mt Vernon, IL 62864
cedarhurst.org

Centralia Foundation Park
600 Pleasant Ave, Centralia, IL 62801
Totality Starts at: 2:00pm CDT
Totality Lasts for: 2:50

Mt Vernon City Park
800 S 27th St
Mt Vernon, IL 62864
www.mtvernon.com

Rend Lake State Waterfowl Management Area
10885 E. Jefferson Rd., Bonnie, IL 62816
Totality Starts at: 2:00pm CDT
Totality Lasts for: 3:47

Sam Dale Lake
612 Co Rd 1910N, Johnsonville, IL 62850
Totality Starts at: 2:01pm CDT
Totality Lasts for: 3:36

Olney, IL
Totality Starts at: 2:02pm CDT
Totality Lasts for: 3:46
Totality Ends at: 2:06pm CDT
NWS Office: www.weather.gov/ilx
City Website: www.ci.olney.il.us

East Fork Lake Campground
5250 E Goosepoint Ln
Olney, IL 62450
www.ci.olney.il.us

Lakeside RV and Campground
5238 N Silver Rd
Olney, IL 62450
Totality Lasts for: 3:46

Newton Lake Conservation Area
3490 E 500th Ave, Newton, IL 62448
Totality Starts at: 2:02pm CDT
Totality Lasts for: 3:05

Olney City Park
502 White Squirrel Cir
Olney, IL 62450
www.ci.olney.il.us

Red Hills State Park
3571 Ranger Lane, Sumner, IL 62466
dnr.illinois.gov/parks/park.redhills.html
Totality Lasts for: 3:58

Rotary Park
5249-5299 E Kiwanis Rd
Olney, IL 62450
www.ci.olney.il.us

Sam Parr State Park
13225 IL-33, Newton, IL 62448
dnr.illinois.gov/parks/park.samparr.html
Totality Lasts for: 3:07

Marshall, IL
Totality Starts at: 2:04pm CDT
Totality Lasts for: 2:38
Totality Ends at: 2:06pm CDT
NWS Office: www.weather.gov/ilx
City Website: marshall-il.com

1918 Brick National Road
US-40, Marshall Township, IL 62441
Totality Starts at: 2:04pm CDT
Totality Lasts for: 2:34

Lincoln Trail State Park
16985 E 1350th Rd
Marshall, IL 62441
dnr.illinois.gov

Mill Creek Lake and Park
20482 N. Park Entrance Rd.
Marshall, IL 62441
www.clarkcountyparkdistrict.com

Never gaze at the sun without eye protection. Only remove eye protection during 100% totality.

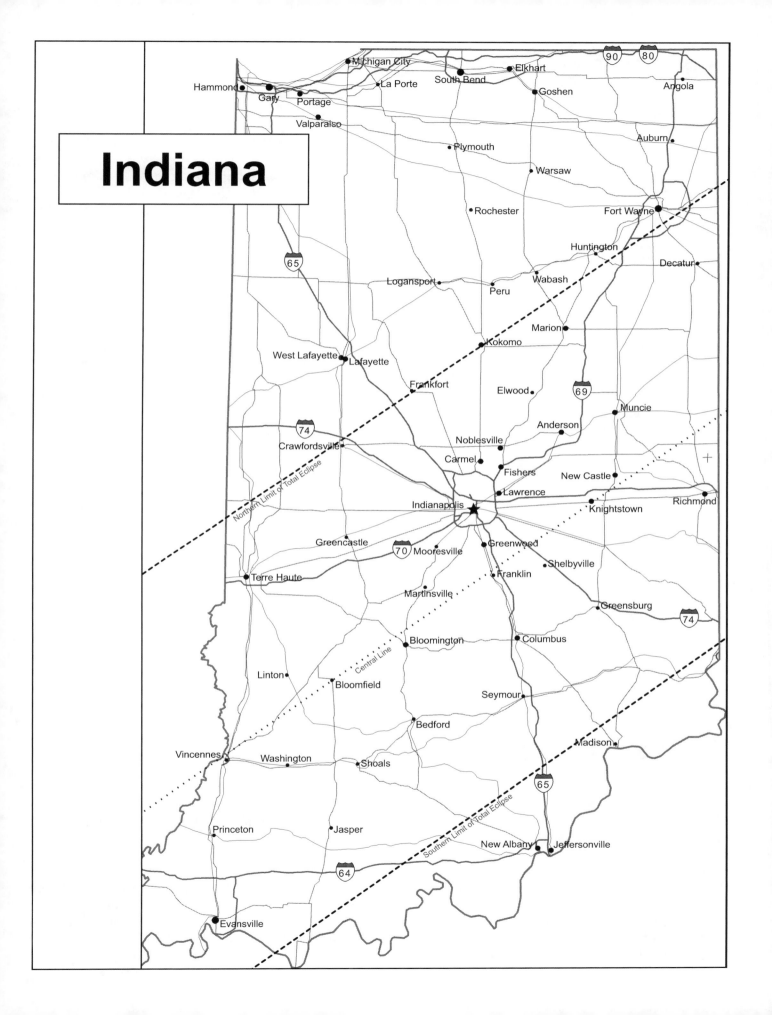

Featured Places to Eat

Babo's Cafe in Terre Haute has a menu filled with Bosnian and Balkan classic dishes. Cevapi features Babo's homemade Bosnian beef sausages served on crispy ciabatta bread with onions and is served with fries. Babo's Cafe also serves an assortment of homemade burgers, such as the Steakhouse Bacon Burger, in addition to sandwiches and wraps. Order your food and brace yourself for 3 minutes of celestial magic in Terre Haute, starting at 3:04pm EDT. Located at 2918 Wabash Ave, (812) 917-0099

Milano's Italian Cuisine in Evansville specializes in classic Italian dishes. Diners can enjoy Lasagna Bolognese, Ziti, Eggplant Parmesana, and other hearty choices. Guests can choose their favorite pasta, and a variety of sauces. Chicken, seafood and pizza are also available. Enjoy over 3 minutes of solar intrigue in Evansville—just remember to fuel up before the dramatic display begins at 2:02pm CDT. Located at 500 Main St, (812) 484-2222

Payne's Restaurant near Marion in Gas City, offers a variety of British cuisine dishes, including classics like Bangers and Mash, British Chicken Curry, Fish 'n Chips, and Beef Stew & Yorkshire pudding. The menu also features grilled dishes, salads, vegetarian and vegan options, breakfast items, appetizers, soups, sandwiches, and ice cream desserts. Fancy a bite before the solar spectacle in Marion? Pop round to this British-themed restaurant and delight in the eclipse, starting at 3:08pm EDT. Located at 4925 S Kay Bee Dr Gas City, (765) 998-0668

Pea-Fections in Vincennes has wide array of delicious sandwiches, salads, and desserts. Sandwiches such as the Cashew Raisin Chicken Salad, and Ham and Swiss are offered, along with many others and sides like potato salad, soup, or fresh fruit. Many salad options are available, like the Pear & Pecan Salad, Greek, Caesar, California Cobb, and the Strawberry Salad. Vincennes is on the eclipse center line so get your food early and enjoy the 4 minute astronomical spectacle! Located at 323 Main St, (812) 886-5146

Shapiro's Delicatessen in Indianapolis has a range of sandwiches, including the Rare Roast Beef Sandwich, which is made with beautifully pink, tender roast beef served cold. Daily specials include Chicken Stew and Split Pea Soup. They also have an astounding array of classic deli sandwiches including favorites such as Pastrami and Corned Beef. A lingering, spell binding totality phase of nearly 4 minutes will happen here. Located at 808 S Meridian St, (317) 631-4041

The Willard in Franklin is focused on sandwiches and pizzas. The Tenderloin sandwiches are billed as the best around, both grilled and breaded varieties. Their tantalizing list of Signature pizzas are available in original or thin crust. They also offer Mexican platters, burgers, and salads. Being near the center of the shadow for this eclipse means over 4 minutes of totality will be on display here. Located at 99 N Main St, (317) 738-9991

Eclipse Track Notes - IN

Cities near the northern and southern limits
Crawfordsville, Frankfort, Kokomo and Fort Wayne are all near the northern limit of the total eclipse. Interstates I-74 , I-65, I-69 all provide a path southward into the path of totality. Kokomo can drop straight down US-31 toward Indianapolis.

Mobility
I-70 cuts across the middle of the eclipse from Terre Haute, to Indianapolis, crosses the centerline near Knightstown, and continues across toward Dayton, OH. I-69 runs from Evansville in the south to Fort Wayne in the north. I-74 crosses the eclipse from Crawfordsville to Indianapolis and toward Cincinnati, Ohio. I-65 provides a north/south route from Layfayette to Indianapolis and south to Louisville, Kentucky. I-64 runs east/ west across the southern tip of Indiana, connecting Evansville, Illinois to Louisville, Kentucky.

Evansville, IN
Totality Starts at: 2:02pm CDT
Totality Lasts for: 3:11
Totality Ends at: 2:05pm CDT
NWS Office: www.weather.gov/pah
City Website: www.evansvillegov.org

Angelo's
305 Main St, (812) 428-6666
Italian, Evansville
Eclipse Day Monday, 11am - 9pm

Biaggi's Ristorante Italiano
6401 E Lloyd Expy #3, (812) 421-0800
Italian, Evansville
Eclipse Day Monday, 11am - 9pm

Casa Fiesta Mexican Restaurant
2121 N Green River Rd Suite 8, (812) 401-4000
Mexican, Evansville
Eclipse Day Monday, 11am - 9pm

Chava's Mexican Grill
4202 N First Ave, (812) 401-1977
Mexican, Evansville
Eclipse Day Monday, 11am - 9pm

Friendship Diner
834 Tutor Ln, (812) 402-0201
American, Evansville
Eclipse Day Monday, 6am - 3pm

Fuji Yama
915 N Park Dr, (812) 962-4440
Japanese, Evansville
Eclipse Day Monday, 11am - 2pm, 4–8:30pm

Gollita Peruvian Cuisine
4313 E Morgan Ave Suite H, (812) 303-5100
Peruvian, Evansville
Eclipse Day Monday, 10:30am - 8pm

Little Italy Restaurant
4430 N First Ave, (812) 401-0588
Italian, Evansville
Eclipse Day Monday, 11am - 9pm

Merry Go Round Restaurant
2101 N Fares Ave, (812) 423-6388
American, Evansville
Eclipse Day Monday, 6:30am - 8:30pm

Milanos Italian Cuisine
500 Main St, (812) 484-2222
Italian, Evansville
Eclipse Day Monday, 11am - 2pm, 4:30–8pm

Vincennes, IN
Totality Starts at: 3:02pm EDT
Totality Lasts for: 4:05
Totality Ends at: 3:06pm EDT
NWS Office: www.weather.gov/ind
City Website: www.vincennes.org

Bill Bobe's Pizzeria
1651 N 6th St, (812) 882-2992
Pizza, Vincennes
Eclipse Day Monday, 11am - 2pm, 4–10pm

Dogwood Barbeque
2232 N 6th St, (812) 882-0552
BBQ, Vincennes
Eclipse Day Monday, 11am - 8pm

Dot's
101 Busseron St, (812) 882-1973
American, Vincennes
Eclipse Day Monday, 7am - 2pm

Olde Thyme Diner
331 Main St, (812) 886-0333
American, Vincennes
Eclipse Day Monday, 7am - 1pm

Pea-Fections
323 Main St, (812) 886-5146
American, Vincennes
Eclipse Day Monday, 10am - 5pm

Senor Tequila
1717 Hart St #5502, (812) 886-3990
Mexican, Vincennes
Eclipse Day Monday, 11am - 9pm

Never gaze at the sun without eye protection. Only remove eye protection during 100% totality.

Terre Haute, IN
Totality Starts at: 3:04pm EDT
Totality Lasts for: 3:03
Totality Ends at: 3:07pm EDT
NWS Office: www.weather.gov/ind
City Website: www.terrehaute.in.gov

Babo's Cafe Inc.
2918 Wabash Ave, (812) 917-0099
Bosnian, Terre Haute
Eclipse Day Monday, 7am - 3pm

Cackleberries
303 S 7th St, (812) 232-0000
American, Terre Haute
Eclipse Day Monday, 6am - 3pm

Coffee Cup
1512 Lafayette Ave, (812) 466-7200
Coffee Shop, Terre Haute
Eclipse Day Monday, 6am - 9pm

Delish Cafe East
8775 Wabash Ave, (812) 877-0001
American, Terre Haute
Eclipse Day Monday, 7am - 2pm

Eastern House
1295 S 3rd St, (812) 234-9898
Chinese, Terre Haute
Eclipse Day Monday, 11am - 9:30pm

First Wok
2570 Wabash Ave, (812) 232-3898
Chinese, Terre Haute
Eclipse Day Monday, 10:30am - 10pm

Greek's Pizzeria
600 Wabash Ave Suite B, (812) 244-1440
Pizza, Terre Haute
Eclipse Day Monday, 11am - 9pm

Lemongrass Thai Restaurant
3830 S US Hwy 41, (812) 814-9299
Thai, Terre Haute
Eclipse Day Monday, 11am - 3pm, 4–8:30pm

M. Moggers Restaurant & Pub
908 Poplar St, (812) 234-9202
American, Terre Haute
Eclipse Day Monday, 11am - 9pm

New Day Cafe
2919 S 3rd St, (812) 235-3200
American, Terre Haute
Eclipse Day Monday, 6am - 9pm

Oy Vey Jewish Bakery and Delicatessen
901 Lafayette Ave, (812) 223-8794
Kosher Restaurant, Terre Haute
Eclipse Day Monday, 9am - 8pm

Pat's Cafe
11890 S US Hwy 41, (812) 299-5637
American, Terre Haute
Eclipse Day Monday, 7am - 2pm

Piloni's Italian Restaurant
1733 Lafayette Ave, (812) 466-4744
Italian, Terre Haute
Eclipse Day Monday, 11am - 9pm

The Bush Family Restaurant
932 Locust St, (812) 238-1148
American, Terre Haute
Eclipse Day Monday, 10am - 8pm

Uncle Jrs BBQ
1429 S 25th St, (812) 814-9050
BBQ, Terre Haute
Eclipse Day Monday, 11am - 10pm

Wise Pies
9 S 6th St, (812) 917-4656
Pizza, Terre Haute
Eclipse Day Monday, 11am - 8pm

Linton, IN
Totality Starts at: 3:03pm EDT
Totality Lasts for: 4:00
Totality Ends at: 3:07pm EDT
NWS Office: www.weather.gov/ind
City Website: www.cityoflinton.com

Casa Sevilla Mexican Restaurant
635 US-231, Bloomfield, (812) 384-3550
Mexican, Bloomfield
Eclipse Day Monday, 11am - 9pm

Cinco De Mayo Mexican
2257 IN-67, Lyons, (812) 659-3600
Mexican, Lyons
Eclipse Day Monday, 11am - 9pm

Cross Roads Cafe
1795 US-231, Bloomfield, (812) 384-4222
American, Bloomfield
Eclipse Day Monday, 6am - 3pm

La Fiesta Mexican
1600 NW A St, (812) 847-0245
Mexican, Linton
Eclipse Day Monday, 11am - 9pm

The Golden Star
180 13th St NW, (812) 512-6050
American, Linton, IN 47441
Eclipse Day Monday, 7am - 3pm

The Grill Inc
60 NW A St, (812) 847-9010
American, Linton,
Eclipse Day Monday, 5am - 7:30pm

Bloomington, IN
Totality Starts at: 3:04pm EDT
Totality Lasts for: 4:02
Totality Ends at: 3:08pm EDT
NWS Office: www.weather.gov/ind
City Website: bloomington.in.gov

Big Woods Bloomington
116 N Grant St, (812) 335-1821
American, Bloomington
Eclipse Day Monday, 11am - 10pm

Blooming Thai
107 N College Ave, (812) 369-4229
Thai, Bloomington
Eclipse Day Monday, 11am - 3pm, 4:30–9pm

Cloverleaf Family Restaurant
4023 W 3rd St, (812) 334-1077
American, Bloomington
Eclipse Day Monday, 5am - 3pm

Cozy Table
2500 W 3rd St, (812) 339-5900
American, Bloomington
Eclipse Day Monday, 6am - 2pm

Dats
408 E 4th St, (812) 339-3090
Cajun, Bloomington
Eclipse Day Monday, 11am - 9pm

Juannita's
620 W Kirkwood Ave, (812) 339-2340
Mexican, Bloomington
Eclipse Day Monday, 11am - 10pm

Little Tibet Restaurant
415 E 4th St, (812) 331-0122
Tibetan, Bloomington
Eclipse Day Monday, 11am - 3pm, 5–9pm

Longfei Chinese Restaurant
113 S Grant St, (812) 955-1666
Chinese, Bloomington
Eclipse Day Monday, 11am - 9:30pm

Mother Bear's Pizza West
2980 W Whitehall Crossing Blvd, (812) 287-7366
Pizza, Bloomington
Eclipse Day Monday, 11am - 9pm

SmokeWorks
121 N College Ave, (812) 287-8190
BBQ, Bloomington
Eclipse Day Monday, 11am - 10pm

Southern Stone
405 W Patterson Dr, (812) 822-3623
American, Bloomington
Eclipse Day Monday, 11am - 9:30pm

The Owlery Restaurant
118 W 6th St, (812) 333-7344
Vegetarian, Bloomington
Eclipse Day Monday, 11am - 9pm

Uptown Cafe
102 E Kirkwood Ave, (812) 339-0900
Cajun & American, Bloomington
Eclipse Day Monday, 8am - 9pm

Mooresville, IN
Totality Starts at: 3:05pm EDT
Totality Lasts for: 3:50
Totality Ends at: 3:09pm EDT
NWS Office: www.weather.gov/ind
City Website: www.mooresville.in.gov

A1 Japanese Steakhouse and Sushi Bar
330 Southbridge St, (317) 831-8883
Japanese, Mooresville
Eclipse Day Monday, 11am - 10pm

Never gaze at the sun without eye protection. Only remove eye protection during 100% totality.

Blueberry Hill Pancake House
460 Town Ctr St, (317) 834-9333
Breakfast, Mooresville
Eclipse Day Monday, 6am - 9pm

Dong's China Buffet
398 S Indiana St, (317) 831-9883
Chinese, Mooresville
Eclipse Day Monday, 11am - 9:30pm

El Rodeo #9
500 Town Ctr St, (317) 831-8753
Mexican, Mooresville
Eclipse Day Monday, 11am - 10pm

Greek's Pizzeria
260 Southbridge St, (317) 483-3200
Pizza, Mooresville
Eclipse Day Monday, 11am - 9pm

Hong Kong Restaurant
340 S Indiana St, (317) 834-3368
Chinese, Mooresville
Eclipse Day Monday, 11am - 8pm

King Gyros
528 S Indiana St, (317) 483-3435
Gyro, Mooresville
Eclipse Day Monday, 10:30am - 10pm

Los Patios
460 S Indiana St, (317) 584-3557
Mexican, Mooresville
Eclipse Day Monday, 11am - 10pm

Morgan's Corner Cafe
457 Town Ctr St, (317) 831-3765
American, Mooresville
Eclipse Day Monday, 6:30am - 2pm

Ralph & Ava's Kitchen & Bar
6 W Main St, (317) 961-4135
American, Mooresville
Eclipse Day Monday, 11am - 10pm

Sal's Famous Pizzeria
360 S Indiana St, (317) 831-0775
Pizza, Mooresville
Eclipse Day Monday, 11am - 9pm

The House
329 Indianapolis Rd, (317) 584-3790
American, Mooresville
Eclipse Day Monday, 8am - 3pm

Indianapolis, IN
Totality Starts at: 3:06pm EDT
Totality Lasts for: 3:47
Totality Ends at: 3:09pm EDT
NWS Office: www.weather.gov/ind
City: www.visitindy.com/eclipse/

317 BBQ
6320 Guilford Ave, (317) 744-0025
BBQ, Indianapolis
Eclipse Day Monday, 11am - 8pm

Barbecue and Bourbon
1414 N Main St, (317) 241-6940
BBQ, Indianapolis
Eclipse Day Monday, 11am - 11pm

Bazbeaux
333 Massachusetts Ave, (317) 636-7662
Pizza, Indianapolis
Eclipse Day Monday, 11am - 9pm

Bluebeard
653 Virginia Ave, (317) 686-1580
American, Indianapolis
Eclipse Day Monday, 11am - 10pm

Daddy Jack's Restaurant & Bar
9419 N Meridian St, (317) 843-1609
American, Indianapolis
Eclipse Day Monday, 11am - 9pm

El Morral Mexican Restaurant LLC
2519 Albany St Beech Grove, (317) 974-9986
Mexican, Indianapolis
Eclipse Day Monday, 11am - 10pm

Flatwater
832 E Westfield Blvd, (317) 257-5466
American, Indianapolis
Eclipse Day Monday, 11am - 10pm

Harry & Izzy's
4050 E 82nd St, (317) 915-8045
Steak, Indianapolis
Eclipse Day Monday, 11am - 10pm

Iozzo's Garden of Italy
946 S Meridian St, (317) 974-1100
Italian, Indianapolis
Eclipse Day Monday, 11am - 2:30pm, 4–9pm

John's Famous Stew
1146 Kentucky Ave, (317) 636-6212
American, Indianapolis
Eclipse Day Monday, 8am - 8pm

King Ribs Bar-B-Q
3145 W 16th St, (317) 488-0223
BBQ, Indianapolis
Eclipse Day Monday, 11am - 9pm

KUMO Japanese Hibachi Steakhouse & Hibachi
1251 U.S. Hwy 31 N P210, (317) 360-6060
Japanese, Greenwood
Eclipse Day Monday, 11am - 3pm, 4–9pm

Loco Mexican Restaurant and Cantina
1417 Prospect St, (317) 384-1745
Mexican, Indianapolis
Eclipse Day Monday, 11am - 9pm

Mimi Blue Restaurants
870 Massachusetts Ave, (317) 737-2625
Meatballs, American, Indianapolis
Eclipse Day Monday, 11am - 10pm

Mug-n-Bun
5211 W 10th St Speedway, (317) 244-5669
American, Speedway
Eclipse Day Monday, 10:30am - 9pm

Nick's Chili Parlor
2621 Lafayette Rd, (317) 924-5005
American, Indianapolis
Eclipse Day Monday, 10:30am - 8pm

Phaya Thai Street Food
5645 N Post Rd, (317) 802-7543
Thai, Indianapolis
Eclipse Day Monday, 11am - 2:30pm, 5–8:30pm

Rackz BBQ
5790 E Main St Suite 140 Carmel, (317) 688-7290
BBQ, Carmel
Eclipse Day Monday, 11am - 8pm

Rick's Cafe Boatyard
4050 Dandy Trail, (317) 290-9300
Seafood, Indianapolis
Eclipse Day Monday, 11am - 10pm

Shapiro's Delicatessen
808 S Meridian St, (317) 631-4041
Deli, Indianapolis
Eclipse Day Monday, 10am - 7:30pm

Sero's Family Restaurant
11720 E Washington St, (317) 894-5570
American, Indianapolis
Eclipse Day Monday, 6am - 3pm

Tios Mexican Restaurant
4863 W Washington St, (317) 426-5206
Mexican, Indianapolis
Eclipse Day Monday, 7am - 7pm

Watami Sushi
1912 Broad Ripple Ave, (317) 991-3355
Sushi, Indianapolis
Eclipse Day Monday, 11am - 2:30pm, 4:30–10

Yats Restaurant
1280 U.S. Hwy 31 N Greenwood, (317) 865-9971
Cajun, Greenwood
Eclipse Day Monday, 11am - 8pm

Franklin, IN
Totality Starts at: 3:05pm EDT
Totality Lasts for: 4:02
Totality Ends at: 3:09pm EDT
NWS Office: www.weather.gov/ind
City Website: www.franklin.in.gov

Ann's Restaurant
77 W Monroe St #2314, (317) 736-5421
American, Franklin
Eclipse Day Monday, 6am - 2pm

Athens
1800 Northwood Plz Dr, (317) 736-8677
Greek, Franklin
Eclipse Day Monday, 11am - 9pm

Dale's 2 Franklin
1071 W Jefferson St, (317) 868-8773
American, Franklin
Eclipse Day Monday, 7am - 3pm

El Pueblo Mexican Restaurant
1904 Northwood Plz Dr, (317) 736-4144
Mexican, Franklin
Eclipse Day Monday, 11am - 10pm

Enzo Pizza
1700 N Morton St, (317) 736-9995
Pizza, Franklin
Eclipse Day Monday, 9:30am - 9pm

Never gaze at the sun without eye protection. Only remove eye protection during 100% totality.

La Cocina Mexican Restaurant
912 N Morton St, (317) 346-0717
Mexican, Franklin
Eclipse Day Monday, 10:30am - 10pm

Mi Abuelito Mexican Restaurant
2797 N Morton St Suite A, (317) 494-6139
Mexican, Franklin
Eclipse Day Monday, 11am - 9pm

The Willard
99 N Main St, (317) 738-9991
American, Franklin
Eclipse Day Monday, 11am - 9pm

Knightstown, IN
Totality Starts at: 3:06pm EDT
Totality Lasts for: 4:01
Totality Ends at: 3:10pm EDT
NWS Office: www.weather.gov/ind
City Website: www.knightstown.in

Gas Grill Family Restaurant
6193 IN-109, (765) 785-6186
American, Knightstown
Eclipse Day Monday, 7am - 9pm

Frosty Boy Restaurant
19 E Main St, (765) 345-5656
Pizza, Knightstown
Eclipse Day Monday, 11am - 8pm

Los Charros D&G Mexican Grill and Bar
15 N Washington St, (765) 571-5268
Mexican, Knightstown
Eclipse Day Monday, 11am - 10pm

Redbone's Pizza & Chicken Co. LLC
16 E Main St, (765) 345-5840
Pizza, Knightstown
Eclipse Day Monday, 11am - 9pm

The Burch Tree Cafe & Bakery
20 N Washington St, (765) 571-5255
American, Knightstown
Eclipse Day Monday, 7am - 3pm

New Castle, IN
Totality Starts at: 3:07pm EDT
Totality Lasts for: 4:00
Totality Ends at: 3:11pm EDT
NWS Office: www.weather.gov/ind
City: www.cityofnewcastle.net

El Chile Poblano Mexican Restaurant
1649 S Memorial Dr, (765) 529-4441
Mexican, New Castle
Eclipse Day Monday, 11am - 10pm

Los Amigos Mexican Restaurant
120 S Memorial Dr, (765) 521-4096
Mexican, New Castle
Eclipse Day Monday, 11am - 9pm

Mancinos Pizzas and Grinders
2111 S Memorial Dr, (765) 529-8868
Pizza, New Castle
Eclipse Day Monday, 11am - 8pm

Park Restaurant
1651 S Memorial Dr, (765) 591-8100
American, New Castle
Eclipse Day Monday, 8am - 8:30pm

Sake Japanese Restaurant Hibachi & Sushi
2201 S Memorial Dr, (765) 388-2660
Japanese, New Castle
Eclipse Day Monday, 11am - 10pm

Stacks Pancake House Restaurant
510 S Memorial Dr, (765) 529-3001
Pancakes, New Castle
Eclipse Day Monday, 8am - 3pm

Anderson, IN
Totality Starts at: 3:07pm EDT
Totality Lasts for: 3:37
Totality Ends at: 3:10pm EDT
NWS Office: www.weather.gov/ind
City: www.cityofanderson.com

Bobber's Cafe at Shadyside Bait & Tackle
1117 Alexandria Pike, (765) 808-5005
American, Anderson
Eclipse Day Monday, 8am - 3pm

Burro Loco
21 W 8th St, (765) 640-6565
Mexican, Anderson
Eclipse Day Monday, 11am - 10pm

Eva's Pancake House
831 Broadway St, (765) 644-9650
Breakfast, Anderson
Eclipse Day Monday, 7am - 3pm

Frazier's Dairy Maid
3311 Main St, (765) 644-5406
Ice Cream, Anderson
Eclipse Day Monday, 11:30am - 9pm

Gene's Root Beer and Hot Dogs
640 S Scatterfield Rd, (765) 642-5768
American, Anderson
Eclipse Day Monday, 11am - 9pm

La Nueva Charreada
1805 University Blvd, (765) 641-2888
Mexican, Anderson
Eclipse Day Monday, 11am - 10pm

Lee's Famous Recipe Chicken
20 E 29th St, (765) 649-0888
American, Anderson
Eclipse Day Monday, 10am - 10pm

Meadowbrook Pizza
5 W 37th St, (765) 644-0929
Pizza, Anderson
Eclipse Day Monday, 11am - 9pm

Riviera Maya Mexica
4434 S Scatterfield Rd, (765) 641-0099
Latin American, Anderson
Eclipse Day Monday, 11am - 10pm

Scampy's Annex
2705 Nichol Ave, (765) 649-7183
Pizza, Anderson
Eclipse Day Monday, 11am - 11pm

The Lemon Drop
1701 Mounds Rd, (765) 644-9055
American, Anderson
Eclipse Day Monday, 10:30am - 9pm

The Toast Cafe
28 E 13th St, (765) 644-8131
American, Anderson
Eclipse Day Monday, 6am - 3pm

Muncie, IN
Totality Starts at: 3:07pm EDT
Totality Lasts for: 3:43
Totality Ends at: 3:11pm EDT
NWS Office: www.weather.gov/ind
City Website: www.cityofmuncie.com

By Hand and Fork
1617 Wheeling Ave, (765) 896-8717
American, Muncie
Eclipse Day Monday, 7am - 3pm

Mulligans Grill
3325 S Walnut St, (765) 282-8129
American, Muncie
Eclipse Day Monday, 8am - 9pm

Puerto Vallarta Mexican Restaurant
3901 N Broadway Ave, (765) 287-8897
Mexican, Muncie
Eclipse Day Monday, 11am - 10pm

Roots Burger Bar
1700 W University Ave, (765) 216-6026
American, Muncie
Eclipse Day Monday, 11am - 10pm

Salsa's Mexican Grill Express
2421 S Madison St, (765) 381-8878
Mexican, Muncie
Eclipse Day Monday, 11am - 9pm

Thai Smile 2 Restaurant
2401 N Tillotson Ave, (765) 289-8989
Thai, Muncie
Eclipse Day Monday, 11am - 2:30pm, 5–8:30pm

Tuppee Tong Thai Restaurant
310 W Main St, (765) 284-3101
Thai, Muncie
Eclipse Day Monday, 11am - 2pm, 5–9pm

Never gaze at the sun without eye protection. Only remove eye protection during 100% totality.

Marion, IN
Totality Starts at: 3:08pm EDT
Totality Lasts for: 2:14
Totality Ends at: 3:10pm EDT
NWS Office: www.weather.gov/iwx
City Website: cityofmarion.in.gov

9th Street Cafe
1802 W 9th St, (765) 664-4851
Breakfast, Marion
Eclipse Day Monday, 5am - 2pm

Esmeralda's Restaurant
2213 Westwood Square, (765) 573-3052
Mexican, Marion
Eclipse Day Monday, 11am - 9pm

Ivanhoes
979 S Main St Upland, (765) 998-7261
American, Marion
Eclipse Day Monday, 10am - 10pm

Los Amores Restaurant
428 S Washington St, (765) 573-6336
Mexican, Marion
Eclipse Day Monday, 11am - 9pm

Myers Drive-In
938 S Washington St, (765) 664-9736
American, Marion
Eclipse Day Monday, 9:30am - 8pm

PC Brick Oven Pizza
2018 W 2nd St, (765) 662-6312
Pizza, Marion
Eclipse Day Monday, 10:30am - 9pm

Payne's Restaurant
4925 S Kay Bee Dr Gas City, (765) 998-0668
British, Marion
Eclipse Day Monday, 11am - 9pm

The Train Station Pancake House
406 E 4th St, (765) 573-4121
American, Marion
Eclipse Day Monday, 6am - 3pm

Richmond, IN
Totality Starts at: 3:07pm EDT
Totality Lasts for: 3:49
Totality Ends at: 3:11pm EDT
NWS Office: www.weather.gov/iln
City: www.richmondindiana.gov

El Bronco North
2515 Chester Blvd, (765) 373-3975
Mexican, Richmond
Eclipse Day Monday, 11am - 9pm

El Rodeo East
3611 E Main St, (765) 965-3340
Mexican, Richmond
Eclipse Day Monday, 11am - 10pm

Firehouse BBQ and Blues
400 N 8th St, (765) 598-5440
BBQ, Richmond
Eclipse Day Monday, 11am - 9pm

Galo's Italian Grill
107 Garwood Rd, (765) 973-9000
Italian, Richmond
Eclipse Day Monday, 11am - 10pm

Gulzar's Indian Cuisine
4712 National Rd E, (765) 939-7401
Indian, Richmond
Eclipse Day Monday, 11am - 2:30pm, 5–9pm

La Mexicana Grill
519 NW 5th St, (765) 965-0561
Mexican, Richmond
Eclipse Day Monday, 11am - 8pm

Lulu's Tacos
4563 National Rd E, (765) 967-4963
Mexican, Richmond
Eclipse Day Monday, 11am - 9pm

Old Richmond Inn
138 S 5th St, (765) 962-2247
American, Richmond
Eclipse Day Monday, 11am - 9pm

Thai Thara
1000 Chester Blvd, (765) 935-1558
Thai, Richmond
Eclipse Day Monday, 11am - 2pm, 5–8pm

Featured Parks and Attractions

The Eiteljorg Museum of American Indians and Western Art is located in downtown Indianapolis, Indiana. It houses one of the world's finest collections of Native contemporary art and Western American paintings and sculptures, with extensive collections from indigenous peoples of the Americas. The museum also features both renowned and contemporary artists such as Remington, O'Keeffe, and Warhol. The eclipse will paint the Indianapolis sky in night-time colors for nearly 4 minutes, be sure to step outside and take it in.

Fort Harrison State Park, also known as Fort Ben, is a day-use park notable for its rich military history and diverse outdoor activities. The park features remnants of a Citizen's Military Training Camp, Civilian Conservation Corps camp, and World War II prisoner of war camp. It also boasts several hiking trails. Visitors can enjoy picnicking, horseback riding, and fishing. The park is home to the Museum of 20th Century Warfare, offering free exhibits on military history and technology. The park is uniquely positioned for the 2024 solar eclipse, providing a rare viewing opportunity of nearly four minutes of totality.

The Indianapolis Motor Speedway Museum is a must-visit destination for motor racing enthusiasts. It houses an impressive collection, including over 25 Indy 500-winning cars, that showcases more than a century of high-speed innovation. A notable highlight is the VIP Basement Collection, formerly accessible only to VIP guests but now open to the public, offering a 30-minute private tour of rare racing artifacts and automobiles. But you don't want to be in the basement around 3pm, because the eclipse shadow will be racing past at over 1,000 miles an hour when it darkens the sky for nearly 4 minutes.

The Indianapolis Zoo offers a unique combination of a zoo, an aquarium, and a botanical garden. Get up close to the animals with immersive experiences such as the Dolphin In-Water Adventure and Animal Art Adventures. Featured attractions include Magnificent Macaws, Kangaroo Crossing, and Race a Cheetah. The zoo is divided into sections such as deserts, oceans, forests, and plains, each one exhibiting an exciting collection of exotic animals.

McCormick's Creek State Park is Indiana's first state park, known for its beautiful waterfalls, canyons, and unique limestone formations. The park offers a variety of amenities including camping, hiking trails, and picnic areas. There's also an inn with lodging and dining options. This park is close to the central line of the eclipse and you'll experience 3:59 of totality here.

Mounds State Park, located in Anderson, Indiana, is known for its rich Native American history. The park is home to ten mounds built by the Adena-Hopewell people over 2000 years ago. Visitors can explore these ancient structures on well-maintained trails. The park also offers amenities such as a nature center, picnic areas, and camping facilities. The eclipse will darken the ancient mounds for over three and a half minutes.

The parks and attractions section in this book are organized by cities, which are highlighted in bold. Under each city, you will find a curated list of locations presented in alphabetical order. These locations may be within the city or situated within a 30-50 mile radius, encompassing the surrounding region. This method of organization allows for easy navigation and planning, ensuring you can make the most of your solar eclipse experience.

Never gaze at the sun without eye protection. Only remove eye protection during 100% totality.

Parks and Attractions

Evansville, IN
Totality Starts at: 2:02pm CDT
Totality Lasts for: 3:11
Totality Ends at: 2:05pm CDT
NWS Office: www.weather.gov/pah
City Website: www.evansvillegov.org

Angel Mounds State Historic Site
8215 Pollack Ave
Evansville, IN 47715
www.indianamuseum.org

Burdette Park
5301 Nurrenbern Rd
Evansville, IN 47712
www.burdettepark.org

Children's Museum of Evansville
22 SE 5th St
Evansville, IN 47708
www.cmoekids.org

Eagle Slough Natural Area
5000 Waterworks Rd
Evansville, IN 47725
sycamorelandtrust.org

Evansville African American Museum
579 S Garvin St
Evansville, IN 47713
www.evvaam.org

Evansville Museum
411 SE Riverside Dr
Evansville, IN 47713
evansvillemuseum.org

Evansville Wartime Museum
7503 Petersburg Rd
Evansville, IN 47725
evansvillewartimemuseum.org

Four Freedoms Monument
201 SE Riverside Dr
Evansville, IN 47713
www.evansvillegov.org

Friedman Park
2700 Park Blvd, Newburgh, IN 47630
www.friedmanpark.com
Totality Lasts for: 2:38

Garvin Park
Evansville, IN 47710
www.visitevansville.com
Totality Lasts for: 3:06

Harmonie State Park Campground
3451 Harmonie State Park Rd, New Harmony
Totality Starts at: 2:01pm CDT
Totality Lasts for: 3:51

Hovey Lake Fish and Wildlife Area
8401 IN-69 S, Mt Vernon, IN 47620
www.in.gov/dnr/fishwild/3092.htm
Totality Lasts for: 3:18

Howell Park
1101 S Barker Ave
Evansville, IN 47712
www.evansvillegov.org

Lincoln State Park
15476 County Rd 300 E, Lincoln City
in.gov/dnr/state-parks/
Totality Lasts for: 2:10

Lynnville Park Campground
405 St. Road 68 West, Lynnville, IN
Totality Starts at: 2:03pm CDT
Totality Lasts for: 3:19

Mesker Park Zoo
1545 Mesker Park Dr
Evansville, IN 47720
www.meskerparkzoo.com

Mickey's Kingdom Park
Riverfront, Downtown
Evansville, IN 47713
www.evansvillegov.org

Moutoux Park
Evansville, IN 47720
www.evansvilleparksfoundation.org
Totality Lasts for: 3:16

Patoka Lake Swimming Beach
County Rd 1100 W, French Lick, IN 47432
Totality Starts at: 2:04pm CDT
Totality Lasts for: 2:43

Reitz Home Museum
112 Chestnut St
Evansville, IN 47713
www.reitzhome.com

Scales Lake Park
800 W Tennyson Rd, Boonville, IN 47601
www.warrickcountyparks.com
Totality Lasts for: 2:39

USS LST-325
610 NW Riverside Dr
Evansville, IN 47708
www.lstmemorial.org

Vann Park
Bayard Park Dr
Evansville, IN 47714
www.evansvillegov.org

Weather Rock Campground
12848 S Weather Rock Dr, Haubstadt
Totality Starts at: 2:02pm CDT
Totality Lasts for: 3:32

Vincennes, IN
Totality Starts at: 3:02pm EDT
Totality Lasts for: 4:05
Totality Ends at: 3:06pm EDT
NWS Office: www.weather.gov/ind
City Website: www.vincennes.org

East Side Park
310 NE 21st St, Washington, IN 47501
Totality Starts at: 3:03pm EDT
Totality Lasts for: 3:58

George Rogers Clark National Historical Park
401 S 2nd St, Vincennes, IN 47591
Totality Starts at: 3:02pm EDT
Totality Lasts for: 4:05

Glendale Fish & Wildlife Area
6001 E 600 S, Montgomery, IN 47558
Totality Starts at: 2:03pm CDT
Totality Lasts for: 3:46

Harrison Mansion Grouseland
3 W Scott St
Vincennes, IN 47591
www.grouseland.org

Indiana Military Museum
715 S 6th St
Vincennes, IN 47591
indianamilitarymuseum.net/visit/

Kimmel Park
Oliphant Dr
Vincennes, IN 47591
www.vincennes.org

Ouabache Trails Campground
Vincennes, IN 47591
www.knoxcountyparks.com
Totality Lasts for: 4:05

The Red Skelton Museum of American Comedy
20 W Red Skelton Blvd
Vincennes, IN 47591
www.redskeltonmuseum.org

White Oak State Fishing Area
4541 N White Oaks Rd, Bruceville, IN 47516
Totality Starts at: 3:02pm EDT
Totality Lasts for: 4:05

Terre Haute, IN
Totality Starts at: 3:04pm EDT
Totality Lasts for: 3:03
Totality Ends at: 3:07pm EDT
NWS Office: www.weather.gov/ind
City Website: terrehaute.in.gov

Bogey's Family Fun Center
3601 Union Rd
Terre Haute, IN 47802
www.bogeybear.net

Cataract Falls State Recreation Area
2605 N Cataract Rd, Spencer, IN 47460
Totality Starts at: 3:04pm EDT
Totality Lasts for: 3:45

Collett Park
2414 N 7th St
Terre Haute, IN 47804
terrehaute.in.gov

Never gaze at the sun without eye protection. Only remove eye protection during 100% totality.

Deming Park
500 S Fruitridge Ave
Terre Haute, IN 47803
terrehaute.in.gov

Dobbs Park Nature Center
5170 E Poplar St
Terre Haute, IN 47803
terrehaute.in.gov

Fairbanks Park
1100 Girl Scout Ln
Terre Haute, IN 47807
terrehaute.in.gov

Hawthorn Park- Vigo County Parks
6067 E Old Maple Ave
Terre Haute, IN 47803
vigoparks.org

Lake Waveland
11677 IN-47, Waveland, IN 47989
Totality Starts at: 3:04pm EDT
Totality Lasts for: 3:38

Lieber State Recreation Area
1317 W Lieber Rd, Cloverdale, IN 46120
Totality Starts at: 3:05pm EDT
Totality Lasts for: 1:20

Maple Avenue Nature Park
500 Maple Ave
Terre Haute, IN 47804
terrehaute.in.gov

Raccoon Lake State Recreation Area
1588 S Raccoon Parkeway, Rockville
www.in.gov/dnr/parklake/2959.ht
Totality Lasts for: 2:35

Terre Haute Children's Museum
727 Wabash Ave
Terre Haute, IN 47807
thchildrensmuseum.com

Linton, IN
Totality Starts at: 3:03pm EDT
Totality Lasts for: 4:00
Totality Ends at: 3:07pm EDT
NWS Office: www.weather.gov/ind
City Website: www.cityoflinton.com

Bloomfield City Park
601-699 W Main St
Bloomfield, IN 47424
www.co.greene.in.us

Eagles Nest Camping
11948 W 200S
Linton, IN 47441
www.eaglesnestcamping.com

Humphrey's Park
1351 E, 1300 NW A St
Linton, IN 47441
www.cityoflinton.com

Redbird State Recreation Area
15470 County Rd 350 N
Dugger, IN 47848
www.in.gov/dnr/state-parks/parks-lakes/

Shakamak State Park
6265 IN-48, Jasonville, IN 47438
in.gov/dnr/state-parks/
Totality Lasts for: 3:48

Sullivan County Park & Lake
990 E Picnic Rd, Sullivan, IN 47882
Totality Starts at: 3:03pm EDT
Totality Lasts for: 3:47

Bloomington, IN
Totality Starts at: 3:04pm EDT
Totality Lasts for: 4:02
Totality Ends at: 3:08pm EDT
NWS Office: www.weather.gov/ind
City Website: bloomington.in.gov

Beanblossom Bottoms Nature Preserve
N Woodall Rd, Ellettsville, IN 47429
Totality Starts at: 3:04pm EDT
Totality Lasts for: 4:03

Bloomington Arboretum
E 10th St
Bloomington, IN 47405
www.visitbloomington.com

Brown County State Park
1801 IN-46, Nashville, IN 47448
in.gov/dnr/state-parks/
Totality Lasts for: 3:57

Burkhart Creek County Park
N Duckworth Rd
Paragon, IN 46166
www.morgancountyparks.org

Captain Janeway Statue
308 W 4th St
Bloomington, IN 47404
janewaycollective.com

Crooked Creek State Recreation Area
T C Steele Rd
Nashville, IN 47448
www.in.gov/dnr/parklake

Eskenazi Museum of Art
1133 E 7th St
Bloomington, IN 47405
artmuseum.indiana.edu

Flatwoods Park
9499 Flatwoods Rd, Gosport, IN 47433
Totality Starts at: 3:04pm EDT
Totality Lasts for: 4:02

Hardin Ridge Recreation Area
6464 Hardin Ridge Rd
Heltonville, IN 47436
www.fs.usda.gov

Hickory Ridge Lookout Tower
Tower Ridge Rd, Norman, IN 47264
Totality Starts at: 3:05pm EDT
Totality Lasts for: 3:50

Karst Farm Park
2450 S Endwright Rd
Bloomington, IN 47403
www.karstfarmpark.com

Leonard Springs Nature Park
4685 S Leonard Springs Rd
Bloomington, IN 47403
bloomington.in.gov

Lower Cascades Park
2851 Old State Rd 37
Bloomington, IN 47404
bloomington.in.gov

McCormick's Creek State Park
250 McCormick Creek Park Rd, Spencer
in.gov/dnr/state-parks/
Totality Lasts for: 3:59

Monroe County History Center
202 E 6th St
Bloomington, IN 47408
monroehistory.org

Morgan-Monroe State Forest
6220 Forest Rd
Martinsville, IN 46151
www.in.gov/dnr/forestry/4816.htm

Paynetown State Recreation Area
4850 State Hwy 446
Bloomington, IN 47401
on.in.gov/monroelake

Pine Grove State Recreation Area
7699 E Pine Grove Rd
Bloomington, IN 47401
www.in.gov/dnr/parklake

Riddle Point Park at Lake Lemon
7599 N Tunnel Rd
Unionville, IN 47468
www.lakelemon.org

Sculpture Trails Outdoor Museum
6764 N Tree Farm Rd, Solsberry, IN 47459
Totality Starts at: 3:04pm EDT
Totality Lasts for: 4:03

Spring Mill State Park
3333 IN-60 E, Mitchell, IN 47446
www.in.gov/dnr/parklake
Totality Lasts for: 3:21

Starve-Hollow State Recreation Area
4345 S 275 W, Vallonia, IN 47281
Totality Starts at: 3:05pm EDT
Totality Lasts for: 3:00

Switchyard Park
1601 S Rogers St
Bloomington, IN 47403
bloomington.in.gov

Never gaze at the sun without eye protection. Only remove eye protection during 100% totality.

Will Detmer Park
4140 W Vernal Pike
Bloomington, IN 47404
www.mcparksandrec.org

WonderLab Science Museum
308 W 4th St
Bloomington, IN 47404
wonderlab.org

Wylie House Museum
307 E 2nd St
Bloomington, IN 47401
libraries.indiana.edu/wylie-house-museum

Mooresville, IN
Totality Starts at: 3:05pm EDT
Totality Lasts for: 3:50
Totality Ends at: 3:09pm EDT
NWS Office: www.weather.gov/ind
City Website: www.mooresville.in.gov

Arbuckle Acres Park
200 N Green St
Brownsburg, IN 46112
brownsburgparks.com/563/Arbuckle-Acres-Park

3-POINT Lake Campgrounds
902 Bunker Hl Rd, Mooresville, IN 46158
Totality Starts at: 3:05pm EDT
Totality Lasts for: 3:52

Hummel Park
1500 S Center St, Plainfield, IN 46168
Totality Starts at: 3:05pm EDT
Totality Lasts for: 3:43

Independence Park
2100 S Morgantown Rd
Greenwood, IN 46143
jocoparks.com

Martinsville City Park
360 N Home Ave, Martinsville, IN 46151
Totality Starts at: 3:05pm EDT
Totality Lasts for: 3:59

McCloud Nature Park
8518 Hughes Rd
North Salem, IN 46165
www.hendrickscountyparks.org

Mooresville Old Town Park
261 E South St #241
Mooresville, IN 46158
www.mooresvillepark.com

Old Town Waverly Park
8425 Main St
Martinsville, IN 46151
www.visitmorgancountyin.com

Pioneer Park
1101 Indianapolis Rd
Mooresville, IN 46158
www.mooresvillepark.com

Quarry Lake Campground
736 S Co Rd 825 E, Fillmore, IN 46128
Totality Starts at: 3:05pm EDT
Totality Lasts for: 3:30

Sodalis Nature Park
7700 S County Rd 975 E
Plainfield, IN 46168
hendrickscountyparks.org

Southwestway Park
8400 Mann Rd
Indianapolis, IN 46221
funfinder.indy.gov/#/details/90

Washington Township Park
115 S County Rd 575 E
Avon, IN 46123
www.washingtontwpparks.org

Indianapolis, IN
Totality Starts at: 3:06pm EDT
Totality Lasts for: 3:47
Totality Ends at: 3:09pm EDT
NWS Office: www.weather.gov/ind
www.visitindy.com/eclipse/

American Legion Mall
700 N Pennsylvania St
Indianapolis, IN 46204
www.in.gov/iwm/2331.htm

Beckenholdt Park
2770 N Franklin St
Greenfield, IN 46140
www.greenfieldin.org

Benjamin Harrison Presidential Site
1230 N Delaware St
Indianapolis, IN 46202
www.presidentbenjaminharrison.org

Brandywine Park
900 E Davis Rd
Greenfield, IN 46140
www.greenfieldin.org

Broad Ripple Park
1426 Broad Ripple Ave
Indianapolis, IN 46220
www.broadripplepark.org

Conner Prairie
13400 Allisonville Rd
Fishers, IN 46038
www.connerprairie.org

Coxhall Gardens
11677 Towne Rd
Carmel, IN 46032
hamiltoncounty.in.gov/1118/Coxhall-Gardens

Depew Memorial Fountain
307 N Meridian St
Indianapolis, IN 46204
wkpd.one/JyzcGU

Eagle Creek Park
7840 W 56th St
Indianapolis, IN 46254
www.indy.gov/activity/about-eagle-creek-park

Eiteljorg Museum
500 W Washington St
Indianapolis, IN 46204
www.eiteljorg.org

Ellenberger Park
5301 E St Clair St
Indianapolis, IN 46219
funfinder.indy.gov/#/details/12

Flat Fork Creek Park
16141 E 101st St
Fishers, IN 46040
www.playfishers.com

Fort Harrison State Park
6000 N Post Rd, Indianapolis, IN 46216
Totality Starts at: 3:06pm EDT
Totality Lasts for: 3:44

Garfield Park
2345 Pagoda Dr
Indianapolis, IN 46203
funfinder.indy.gov/#/details/134

Holliday Park
6363 Spring Mill Rd
Indianapolis, IN 46260
www.hollidaypark.org

Indiana State Museum
650 W Washington St
Indianapolis, IN 46204
www.indianamuseum.org

Indianapolis Fire Fighters Museum
748 Massachusetts Ave
Indianapolis, IN 46204
L416.com

Indianapolis KOA Holiday
5896 W 200 N
Greenfield, IN 46140
koa.com/campgrounds/indianapolis/

Indianapolis Motor Speedway Museum
4750 W 16th St, Indianapolis, IN 46222
Heritage Museum
imsmuseum.org

Indianapolis Zoo
1200 W Washington St
Indianapolis, (317) 630-2001
www.indianapoliszoo.com

Morse Park & Beach
19777 Morse Park Ln
Noblesville, IN 46060
www.hamiltoncounty.in.gov

Mulberry Fields
9645 Whitestown Rd
Zionsville, IN 46077
www.zionsville-in.gov

Northwestway Park
5253 W 62nd St
Indianapolis, IN 46268
funfinder.indy.gov/#/details/32

Oldfields–Lilly House & Gardens
4000 N Michigan Rd
Indianapolis, IN 46208
www.imamuseum.org

Never gaze at the sun without eye protection. Only remove eye protection during 100% totality.

Riverside Park
2420 E Riverside Dr
Indianapolis, IN 46208
funfinder.indy.gov/#/details/135

Soldiers & Sailors Monument
1 Monument Cir, Indianapolis, IN 46204
Observation Deck
www.in.gov/iwm/

Southeastway Park
5624 S Carroll Rd
New Palestine, IN 46163
funfinder.indy.gov/#/details/27

Strawtown Koteewi Park
12308 Strawtown Ave
Noblesville, IN 46060
www.hamiltoncounty.in.gov

The Children's Museum of Indianapolis
3000 N Meridian St
Indianapolis, IN 46208
www.childrensmuseum.org

USS Indianapolis National Memorial
Canal Walk
Indianapolis, IN 46202
www.indianawarmemorials.org

Virginia B. Fairbanks Art & Nature Park
1850 W 38th St
Indianapolis, IN 46228
discovernewfields.org

White River State Park
801 W Washington St
Indianapolis, IN 46204
whiteriverstatepark.org

Franklin, IN
Totality Starts at: 3:05pm EDT
Totality Lasts for: 4:02
Totality Ends at: 3:09pm EDT
NWS Office: www.weather.gov/ind
City Website: www.franklin.in.gov

Beaver Bottom Lake
3425 County Rd 550 S
Franklin, IN 46131
Totality Lasts for: 4:00

Blue Heron Park
405 Driftwood Ct
Franklin, IN 46131
www.franklin.in.gov

Blue River Memorial Park
725 Lee Blvd, Shelbyville, IN 46176
Totality Starts at: 3:06pm EDT
Totality Lasts for: 3:59

Clifty Park
Indiana Ave & Marr Road, Columbus, IN 47201
Totality Starts at: 3:06pm EDT
Totality Lasts for: 3:42

Johnson County Museum of History
135 N Main St
Franklin, IN 46131
johnsoncountymuseum.org

Johnson County Park
2949 E North St
Nineveh, IN 46164
jocoparks.com/johnson-county-park/

Mill Race Park
50 Carl Miske Drive, Columbus, IN 47201
Totality Starts at: 3:05pm EDT
Totality Lasts for: 3:46

Old City Park
304 S Meridian St
Greenwood, IN 46142
www.greenwood.in.gov

Palmer Park
434-450 W Madison St
Franklin, IN 46131
www.franklin.in.gov

Proctor Park
499 Tracy Rd
Whiteland, IN 46184
newwhiteland.in.gov

New Castle, IN
Totality Starts at: 3:07pm EDT
Totality Lasts for: 4:00
Totality Ends at: 3:11pm EDT
NWS Office: www.weather.gov/ind
www.cityofnewcastle.net

Baker Park
New Castle
IN 47362
www.cityofnewcastle.net

Henry County Memorial Park
260 W County Rd 100 N
New Castle, IN 47362
www.henryco.net

Province Pond
N County Rd 125 W
New Castle, IN 47362
www.cityofnewcastle.net

Summit Lake State Park
5993 N Messick Rd, New Castle, IN 47362
in.gov/dnr/state-parks/parks-lakes/
Totality Lasts for: 3:56

Sunset Park & Splash Pad
206 S Hill Ave
Knightstown, IN 46148
www.knightstown.in

Westwood Park
1900 S County Rd 275 W
New Castle, IN 47362
www.visitwestwood.com

Anderson, IN
Totality Starts at: 3:07pm EDT
Totality Lasts for: 3:37
Totality Ends at: 3:10pm EDT
NWS Office: www.weather.gov/ind
www.cityofanderson.com

Anderson Museum of Art
32 Hst W 10th St
Anderson, IN 46016
www.andersonart.org

Anderson Speedway
1311 Martin Lthr Kng Jr Blvd
Anderson, IN 46016
andersonspeedway.com

Falls Park
460 Falls Park Dr
Pendleton, IN 46064
www.fallspark.org

Mounds State Park Family Campgrounds
Anderson, IN 46017
Totality Starts at: 3:07pm EDT
Totality Lasts for: 3:41

Mystic Waters Campground
5435 W State Rd 38
Pendleton, IN 46064
mysticwaterscampground.com

Rangeline Nature Preserve
1200 S Rangeline Rd
Anderson, IN 46012
www.cityofanderson.com

River Bend Park
46012, 201 Sycamore St
Anderson, IN 46016
www.cityofanderson.com

Shadyside Memorial Park Chinese Gardens
1112 Broadway St
Anderson, IN 46012
www.cityofanderson.com

Worlds Largest Ball Of Paint
10696 N 200 W
Alexandria, IN 46001
ballofpaint.freehosting.net

Muncie, IN
Totality Starts at: 3:07pm EDT
Totality Lasts for: 3:43
Totality Ends at: 3:11pm EDT
NWS Office: www.weather.gov/ind
City Website: www.cityofmuncie.com

Cardinal Greenway Muncie Depot Trailhead
700 E Wysor St
Muncie, IN 47305
cardinalgreenways.org

Never gaze at the sun without eye protection. Only remove eye protection during 100% totality.

Charles W. Brown Planetarium
2111 W Riverside Ave
Muncie, IN 47306
www.bsu.edu/planetarium

David Owsley Museum of Art
2021 W Riverside Ave
Muncie, IN 47306
www.bsu.edu/doma

Heekin Park
1797 S Hackley St
Muncie, IN 47302
www.cityofmuncie.com

Muncie Children's Museum
515 S High St
Muncie, IN 47305
www.munciemuseum.com

National Model Aviation Museum
5151 E Memorial Dr
Muncie, IN 47302
www.modelaircraft.org/museum

Oakhurst Gardens
1200 N Minnetrista Pkwy
Muncie, IN 47303
www.cityofmuncie.com

The Bob Ross Experience
620 W Minnetrista Blvd
Muncie, IN 47303
www.minnetrista.net/bobrossexperience

Tuhey Park
W White River Blvd & North Wheeling Ave
Muncie, IN 47303
www.cityofmuncie.com

White River Park
Muncie, IN 47303
Totality Starts at: 3:07pm EDT
Totality Lasts for: 3:43

Marion, IN
Totality Starts at: 3:08pm EDT
Totality Lasts for: 2:14
Totality Ends at: 3:10pm EDT
NWS Office: www.weather.gov/iwx
City Website: cityofmarion.in.gov

Gas City Park
701 S Broadway St
Gas City, IN 46933
www.showmegrantcounty.com

Hostess House (J. Wood Wilson House)
723 W 4th St
Marion, IN 46952
www.hostesshouse.org

James Dean Gallery
425 N Main St
Fairmount, IN 46928
www.jamesdeangallery.com

Jonesboro Community Park
708 Fairmount Ave
Jonesboro, IN 46938
Totality Lasts for: 2:43

Matter Park
N River Rd &, N Quarry Rd
Marion, IN 46952
www.showmegrantcounty.com/places/matter-park/

Quilters Hall of Fame (Webster House)
926 S Washington St
Marion, IN 46953
quiltershalloffame.net

The James Dean Museum
203 E Washington St
Fairmount, IN 46928
www.thejamesdeanmuseum.com/

Richmond, IN
Totality Starts at: 3:07pm EDT
Totality Lasts for: 3:49

Totality Ends at: 3:11pm EDT
NWS Office: www.weather.gov/iln
www.richmondindiana.gov

Clear Creek Park
1201 W Main St
Richmond, IN 47374
www.richmondindiana.gov

Glen Miller Park
2200 E Main St
Richmond, IN 47374
www.richmondindiana.gov

Hayes Regional Arboretum
801 Elks Rd
Richmond, IN 47374
www.hayesarboretum.org

Historic Connersville
200 W 5th St
Connersville, IN 47331
historicconnersville.org

Jack Elstro Plaza
47 N 6th St
Richmond, IN 47374
www.richmondindiana.gov

Joseph Moore Museum
801 W Nat'l Rd
Richmond, IN 47374
earlham.edu/jmm

Middlefork Reservoir
1750 Sylvan Nook Dr
Richmond, IN 47374
www.richmondindiana.gov

Model T Museum
309, 310 N 8th St
Richmond, IN 47374
www.mtfca.com/museum

Richmond KOA Holiday
3101 Cart Rd
Richmond, IN 47374
koa.com/campgrounds/richmond/

Richmond Parks and Recreation
2200 E Main St
Richmond, IN 47374
www.richmondindiana.gov

Richmond Rose Garden
Lower Dr
Richmond, IN 47374
richmondrosegarden.com

Springwood Park
Waterfall Rd
Richmond, IN 47374
www.richmondindiana.gov

Thistlethwaite Falls
65 Waterfall Rd
Richmond, IN 47374
www.richmondindiana.gov

Veterans Memorial Park
Cardinal Greenway
Richmond, IN 47374
www.veteransmemorialpark.org

Wayne County Historical Museum
1150 N A St
Richmond, IN 47374
wchmuseum.org

Whitewater Memorial State Park
1418 S State Rd 101, Liberty, IN 47353
Totality Starts at: 3:07pm EDT
Totality Lasts for: 3:32

Never gaze at the sun without eye protection. Only remove eye protection during 100% totality.

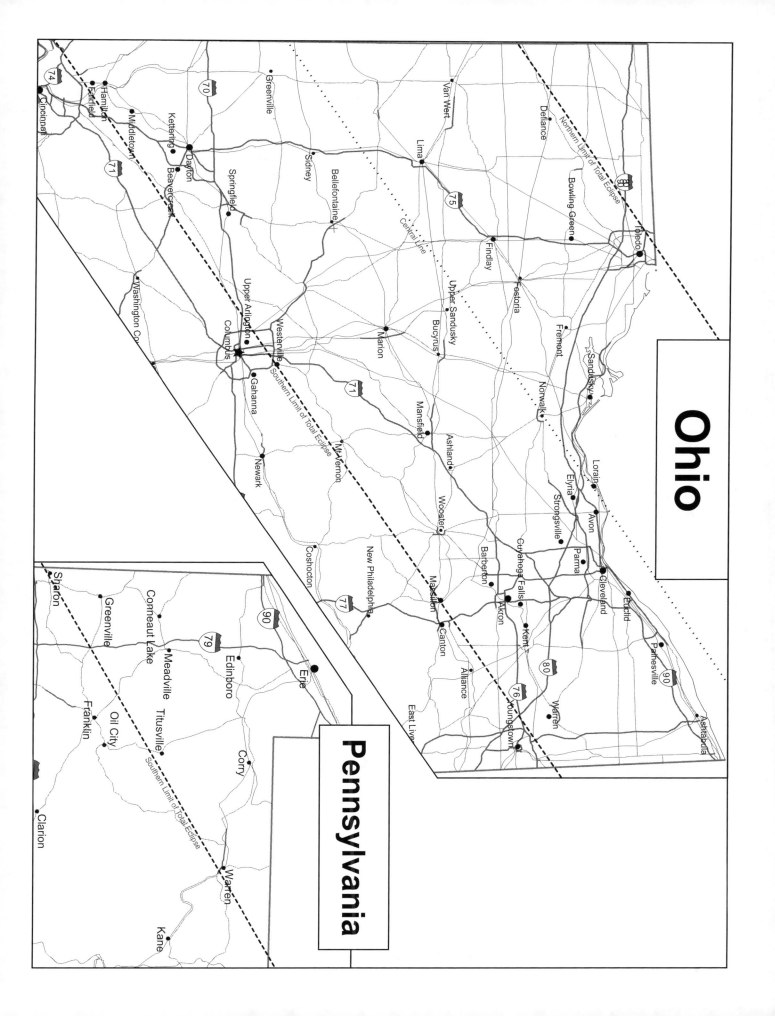

Featured Places to Eat

Belgrade Gardens in Barberton has been serving Famous Chicken Dinners since 1933. Hand-breaded Crisp, Golden Fried Chicken or Chicken Paprikash stewed in a spicy sauce with dumplings are hearty meals. Check out the pork and beef options such as the Balkan Grill (cevap) and Grilled Pork Chops. Secure your chicken and bask in 2:41 of solar eclipse splendor in Barberton, beginning at 3:14pm. Located at 401 E State St, (330) 745-0113

Chemtag African Kitchen in Dayton features African cuisine in a casual setting. A popular menu item is Jollof Rice, which is made with rice, tomatoes, onions, peppers, and a meat like fish or chicken. One of the many other dishes they offer is Beef Spinach Stew with Fufu, a traditional West African staple food made from starchy vegetables processed into a dough. You'll be delighted to dine and see 3 minutes of the total solar eclipse from here. Located at 5254 N Dixie Dr, (937) 430-1911

Scooter's World Famous Dawg House near Painesville in Mentor boasts an array of savory Dawg House Specialties and satisfying meals. They have creatively-topped hot dogs, such as the Mac & Cheese Dawg, topped with shells and Monterey jack cheese. Scooter's also offers a range of hearty sandwiches and ice cream. Get your food and settle in for almost 4 minutes of sun-moon harmony in Mentor, starting at 3:14pm. Located at 9600 Blackbrook Rd, Mentor, (440) 354-8480

The Fern in Findlay has hearty dinner options like the Legendary Meatloaf Dinner and the savory Smoked Pulled Pork Dinner. Sandwich enthusiasts can enjoy the classic Gyro, the Shrimp PO' Boy, or one of the restaurant's signature Legendary Meatloaf Sandwiches. The menu also features satisfying wraps and salads. Totality takes over for nearly 4 minutes in Findlay at 3:10pm. Don't forget to order your food early and witness the remarkable solar event. Located at 452 E Sandusky St, (419) 423-2700

Tony Packo's in Toledo offers a menu that celebrates Hungarian-American cuisine. The restaurant is famous for its Packo's Original Hot Dog, a Hungarian-style hot dog served with mustard, onions, and a special hot dog sauce. Specialties include Chicken Paprikas, Stuffed Cabbage, and a variety of sandwiches. Pierogies are another menu highlight, with options like Chili & Cheese Pierogies. The sun and moon align for 2 minutes in Toledo. Get your meal and prepare to be enchanted, starting at 3:12pm. Located at 1902 Front St, (419) 691-6054

Tony's Family Restaurant near Cleveland in Parma offers a delightful lunch menu with Open Face Sandwiches including Roast Turkey and Pot Roast, as well as Fresh Wraps like the Honey Dijon Chicken Wrap. Their Panini Pressed Sandwiches such as the Ultimate Turkey Panini and the Spicy Chorizo Chicken Panini are hot and delicious. They also have many delicious burgers, sandwiches, and homemade Pierogies. The sky goes dark for almost 4 minutes in Cleveland as the eclipse commences; make sure you're fed and ready to witness the spectacle! Located at 1515 W Pleasant Valley Rd, (440) 842-2250

Eclipse Track Notes - OH - PA

Cities near the northern and southern limits
Cincinnati, Columbus, Canton, and Youngstown are all near the southern limit of the total eclipse. I-75, I-71, and I-77 all provide a route north into the path of totality. Youngstown can take US-422 up to Warren.

Mobility
Transit wise, I-75 from Dayton in the south to Toledo in the north, I-71 from Columbus to Cleveland, and I-77 from Canton to Cleveland all provide north/south routes in the path of totality. US-30 runs east/west across the eclipse from Fort Wayne, Indiana to Wooster, Ohio. I-70 runs east/west above Dayton, connecting to Columbus. In the north along Lake Erie, I-90 connects Toledo, Cleveland and Erie, Pennsylvania. In Pennsylvania, I-79 provides a direct route into the path of totality from Pittsburgh up to Erie.

Greenville, OH
Totality Starts at: 3:08pm EDT
Totality Lasts for: 3:56
Totality Ends at: 3:12pm EDT
NWS Office: www.weather.gov/iln
City: www.cityofgreenville.org

Coles Front Street Inn Restaurant
812 Front St, (937) 548-8727
American, Greenville
Eclipse Day Monday, 5am - 2pm

Dairy Barn
1271 Sweitzer St, (937) 548-3555
American, Greenville
Eclipse Day Monday, 11am - 9pm

Las Marias
1160 Russ Rd, (937) 316-8484
Mexican, Greenville
Eclipse Day Monday, 11am - 10pm

Maid-Rite Sandwich Shoppe
125 N Broadway St, (937) 548-9340
American, Greenville
Eclipse Day Monday, 11am - 8pm

Montage
527 S Broadway St, (937) 548-1950
American, Greenville
Eclipse Day Monday, Open 24 hours

The Coffee Pot
537 S Broadway St #101, (937) 459-5498
Coffee Shop, Greenville
Eclipse Day Monday, 6:30am - 6pm

Dayton, OH
Totality Starts at: 3:09pm EDT
Totality Lasts for: 3:00
Totality Ends at: 3:12pm EDT
NWS Office: www.weather.gov/iln
City Website: www.daytonohio.gov

Abner's Restaurant
2424 E 3rd St, (937) 252-3318
American, Dayton
Eclipse Day Monday, 8am - 4pm

Benjamin's the Burger Master
1000 N Main St, (937) 223-8702
American, Dayton
Eclipse Day Monday, 11am - 7pm

Burkey Family Restaurant
670 Shiloh Springs Rd, (937) 275-7127
American, Dayton
Eclipse Day Monday, 8am - 7:30pm

Chemtag African Kitchen
5254 N Dixie Dr, (937) 430-1911
African, Dayton
Eclipse Day Monday, 10am - 8pm

City Barbeque
6549 Miller Ln, (937) 200-1660
BBQ, Dayton
Eclipse Day Monday, 11am - 9pm

Dayton Village Pizza HALAL Turkish
3630 N Dixie Dr, (937) 567-0775
Mediterranean, Dayton
Eclipse Day Monday, 11am - 9pm

Debbie's Restaurant
2620 Valley Pike, (937) 236-2220
American, Dayton
Eclipse Day Monday, 6am - 8pm

Dewey's Pizza
131 Jasper St, (937) 223-0000
Pizza, Dayton
Eclipse Day Monday, 11am - 9pm

Flying Pizza
223 N Main St, (937) 222-8031
Pizza, Dayton
Eclipse Day Monday, 10am - 7pm

Istanbul Grill
2021 Republic Dr, (937) 387-9504
Turkish, Dayton
Eclipse Day Monday, 11am - 10pm

Legacy Pancake House
1510 N Keowee St, (937) 222-2037
Breakfast, Dayton
Eclipse Day Monday, 7am - 2:30pm

Marion's Piazza
3443 N Dixie Dr, (937) 277-6553
Pizza, Dayton
Eclipse Day Monday, 11am - 10pm

Never gaze at the sun without eye protection. Only remove eye protection during 100% totality.

Olive Mediterranean Grill
44 W Third St, (937) 221-8399
Mediterranean, Dayton
Eclipse Day Monday, 11am - 2pm, 5–8pm

Shen's Szechuan & Sushi
7580 Poe Ave, (937) 898-3860
Chinese, Dayton
Eclipse Day Monday, 11am - 10pm

Smokey Bones Dayton
6744 Miller Ln, (937) 415-0185
BBQ, Dayton
Eclipse Day Monday, 11am - 1am

Smokin Bar-B-Que
200 E 5th St, (937) 586-9790
BBQ, Dayton
Eclipse Day Monday, 10:30am - 9pm

Submarine House
7850 N Main St, (937) 898-9117
Sandwich, Dayton
Eclipse Day Monday, 11am - 9pm

Taqueria Mixteca
1609 E 3rd St, (937) 258-2654
Mexican, Dayton
Eclipse Day Monday, 10am - 9pm

Victor's Taco Shop
1438 N Keowee St, (937) 224-3293
Mexican, Dayton
Eclipse Day Monday, 9am - 10pm

Sidney, OH
Totality Starts at: 3:09pm EDT
Totality Lasts for: 3:53
Totality Ends at: 3:13pm EDT
NWS Office: www.weather.gov/iln
City Website: www.sidneyoh.com

Alcove Restaurant
134 N Main Ave, (937) 492-3737
American, Sidney
Eclipse Day Monday, 5:30am - 3pm

Cassano's Pizza & Subs
1294 Wapakoneta Ave, (937) 492-3115
Pizza, Sidney
Eclipse Day Monday, 11am - 10pm

Clancy's Hamburgers
1250 Wapakoneta Ave, (937) 492-8820
American, Sidney
Eclipse Day Monday, 6am - 9pm

Murphy's Craftbar + Kitchen
110 E Poplar St, (937) 658-6160
American, Sidney
Eclipse Day Monday, 11am - 10pm

Smok'n Jo's BBQ
1951 W Michigan St, (937) 710-4076
BBQ, Sidney
Eclipse Day Monday, 11am - 9pm

Spot Restaurant
201 S Ohio Ave, (937) 492-9181
American, Sidney
Eclipse Day Monday, 11am - 7pm

Lima, OH
Totality Starts at: 3:09pm EDT
Totality Lasts for: 3:49
Totality Ends at: 3:13pm EDT
NWS Office: www.weather.gov/iwx
City Website: www.limaohio.com

Beer Barrel Pizza & Grill
2755 Harding Hwy Suite B, (419) 229-6211
Pizza, Lima
Eclipse Day Monday, 11am - 11pm

Happy Daz
1064 Bellefontaine Ave, (419) 227-3663
American, Lima
Eclipse Day Monday, 8am - 9pm

Joey's Subs
124 N Main St, (419) 224-5639
American, Lima
Eclipse Day Monday, 10am - 6pm

LuLu's Diner
2114 Spencerville Rd, (419) 221-5858
American, Lima
Eclipse Day Monday, 6am - 2pm

Findlay, OH
Totality Starts at: 3:10pm EDT
Totality Lasts for: 3.42
Totality Ends at: 3:14pm EDT
NWS Office: www.weather.gov/cle
City Website: www.findlayohio.gov

Circle of Friends Restaurant
125 W Sandusky St, (567) 294-4274
Asian, Findlay
Eclipse Day Monday, 11am - 9pm

Fern Cafe
452 E Sandusky St, (419) 423-2700
American, Findlay
Eclipse Day Monday, 10am - 11pm

Golden Flames Asian Bistro Express
2033 Tiffin Ave Suite 2, (567) 525-5094
Asian, Findlay
Eclipse Day Monday, 10:30am - 8:30pm

Jack-B's
517 W Trenton Ave, (567) 294-4234
American, Findlay
Eclipse Day Monday, 11am - 9pm

La Charrita
3210 N Main St, (419) 422-8226
Mexican, Findlay
Eclipse Day Monday, 11am - 9pm

Oler's Bar & Grill
708 Lima Ave, (419) 423-2846
Mexican, Findlay
Eclipse Day Monday, 11am - 10pm

Pilgrim Family Restaurant
1505 W Main Cross St, (419) 422-7022
American, Findlay
Eclipse Day Monday, 6am - 2pm

Preteroti's Spaghetti House
1331 N Main St, (567) 250-9944
Italian, Findlay
Eclipse Day Monday, 11am - 9pm

Rossilli's Restaurant
217 S Main St, (419) 423-5050
Italian, Findlay
Eclipse Day Monday, 11am - 2pm, 5–9pm

Bowling Green, OH
Totality Starts at: 3:11pm EDT
Totality Lasts for: 3:00
Totality Ends at: 3:14pm EDT
NWS Office: www.weather.gov/cle
City Website: www.bgohio.org

Beckett's Burger Bar & Barrel Room
163 S Main St, (419) 352-7800
American, Bowling Green
Eclipse Day Monday, 11am - 10pm

Easy Street Cafe
104 S Main St, (419) 353-0988
American, Bowling Green
Eclipse Day Monday, 11am - 9pm

El Zarape
1616 E Wooster St # 1, (419) 353-0937
Mexican, Bowling Green
Eclipse Day Monday, 11am - 9pm

Guajillo's Cocina Mexicana
434 E Wooster St C, (419) 806-4866
Mexican, Bowling Green
Eclipse Day Monday, 11am - 8:30pm

Kabob It Bowling Green
132 E Wooster St, (567) 413-4700
Mediterranean, Bowling Green
Eclipse Day Monday, 11am - 10pm

Kermit's Family Restaurant
307 S Main St, (419) 354-1388
American, Bowling Green
Eclipse Day Monday, 6:30am - 2:30pm

Mr. Spots
206 N Main St, (419) 352-7768
Sandwich, Bowling Green
Eclipse Day Monday, 11am - 9pm

SamB's Restaurant
146 N Main St, (419) 353-2277
American, Bowling Green
Eclipse Day Monday, 11am - 9pm

Never gaze at the sun without eye protection. Only remove eye protection during 100% totality.

Toledo, OH
Totality Starts at: 3:12pm EDT
Totality Lasts for: 2:00
Totality Ends at: 3:14pm EDT
NWS Office: www.weather.gov/cle
City Website: toledo.oh.gov

American Table Family Restaurant
846 S Wheeling St, Oregon, (419) 690-7230
American, Oregon
Eclipse Day Monday, 7am - 8pm

Carlos Poco Loco
1809 Adams St, (419) 214-1655
Cuban & Mexican, Toledo
Eclipse Day Monday, 11am - 9pm

El Camino Real Sky
2072 Woodville Rd Oregon, (419) 693-6695
Mexican, Toledo
Eclipse Day Monday, 11am - 9pm

Grumpy's Toledo
34 S Huron St, (419) 241-6728
American, Toledo
Eclipse Day Monday, 10am - 2pm

Luckie's Barn & Grill
3310 Navarre Ave Oregon, (419) 725-4747
American, Toledo
Eclipse Day Monday, 11am - 11pm

Tony Packo's Restaurant
1902 Front St, (419) 691-6054
American, Toledo
Eclipse Day Monday, 10:30am - 10pm

Woodville Diner
1949 Woodville Rd Oregon, (419) 691-9999
American, Oregon
Eclipse Day Monday, 7am - 8pm

Upper Sandusky, OH
Totality Starts at: 3:11pm EDT
Totality Lasts for: 3:55
Totality Ends at: 3:14pm EDT
NWS Office: www.weather.gov/cle
City: www.uppersanduskyoh.com

A.J.'s Heavenly Pizza
131 N Sandusky Ave, (419) 209-0095
Pizza, Upper Sandusky
Eclipse Day Monday, 11am - 8pm

Cheers 2 U
119 N Sandusky Ave, (419) 294-0900
BBQ, Upper Sandusky
Eclipse Day Monday, 11am - 9pm

Corner Inn
143 N Sandusky Ave, (419) 294-5201
American, Upper Sandusky
Eclipse Day Monday, 6am - 2pm

Los Arcos
370 N Warpole St, (419) 294-9100
Mexican, Upper Sandusky
Eclipse Day Monday, 11am - 9pm

The Village Restaurant
435 N Warpole St, (419) 294-2945
American, Upper Sandusky
Eclipse Day Monday, 11am - 4pm

Mansfield, OH
Totality Starts at: 3:12pm EDT
Totality Lasts for: 3:17
Totality Ends at: 3:15pm EDT
NWS Office: www.weather.gov/cle
City Website: ci.mansfield.oh.us

Athens Greek Restaurant
43 N Lexington-Springmill Rd, (419) 589-4976
Greek, Ontario
Eclipse Day Monday, 11am - 9pm

Los 3 Mayas
595 Ashland Rd, (419) 589-0017
Mexican, Mansfield
Eclipse Day Monday, 11am - 10pm

Mansfield Family Restaurant- Southside
948 S Main St, (419) 756-0479
American, Mansfield
Eclipse Day Monday, 7am - 8pm

Porky's Drive In
811 Ashland Rd, (419) 589-9933
American, Mansfield
Eclipse Day Monday, 10am - 8:45pm

Rancho Fiesta
1360 S Trimble Rd, (419) 774-1744
Mexican, Mansfield
Eclipse Day Monday, 11am - 10pm

Uncle John's Place
18 S Main St, (419) 526-9197
American, Mansfield
Eclipse Day Monday, 11am - 9pm

Norwalk, OH
Totality Starts at: 3:12pm EDT
Totality Lasts for: 3:54
Totality Ends at: 3:16pm EDT
NWS Office: www.weather.gov/cle
City Website: www.norwalkoh.com

Casa Bravos
203 Cline St, (567) 743-9123
Mexican, Norwalk
Eclipse Day Monday, 11am - 10pm

Casa Fiesta
344 Milan Ave, (419) 660-8085
Mexican, Norwalk
Eclipse Day Monday, 11am - 10pm

Jimmy's Backyard BBQ
201 Milan Ave, (419) 668-2200
BBQ, Norwalk
Eclipse Day Monday, 11am - 9pm

Sheri's Coffee House
27 Whittlesey Ave, (419) 663-5282
Coffee Shop, Norwalk
Eclipse Day Monday, 7am - 6pm

The Freight House Pub & Grill
50 N Prospect St, (567) 424-6573
American, Norwalk
Eclipse Day Monday, 11am - 8pm

The Press Box
21 Mill St, (419) 660-7014
American, Norwalk
Eclipse Day Monday, 11am - 9:30pm

Avon, OH
Totality Starts at: 3:13pm EDT
Totality Lasts for: 3:53
Totality Ends at: 3:17pm EDT
NWS Office: www.weather.gov/cle
City Website: www.cityofavon.com

Hecks Of Avon
35514 Detroit Rd, (440) 937-3200
American, Avon
Eclipse Day Monday, 11am - 9pm

Moe's Southwest Grill
36050 Detroit Rd, (440) 934-5663
Mexican, Avon
Eclipse Day Monday, 11am - 8pm

Swensons Drive-In
36041 Main St, (440) 493-1934
American, Avon
Eclipse Day Monday, 11am - 12am

Cleveland, OH
Totality Starts at: 3:13pm EDT
Totality Lasts for: 3:50
Totality Ends at: 3:17pm EDT
NWS Office: www.weather.gov/cle
City: www.thisiscleveland.com

Addy's Diner
530 Euclid Ave, (216) 202-1368
American, Cleveland
Eclipse Day Monday, 6:30am - 2pm

Adam's Place Restaurant
681 E 200th St Euclid, (216) 531-9681
American, Cleveland
Eclipse Day Monday, 7am - 2:30pm

Angelo's Pizza
13715 Madison Ave Lakewood, (216) 221-0440
Pizza, Cleveland
Eclipse Day Monday, 11am - 11pm

Never gaze at the sun without eye protection. Only remove eye protection during 100% totality.

Arrabiata's Italian Restaurant
6169 Mayfield Rd, (440) 442-2600
Italian, Cleveland
Eclipse Day Monday, 11:30am - 9pm

Cozumel Mexican Restaurant
4195 W 150th St, (216) 331-4310
Mexican, Cleveland
Eclipse Day Monday, 11am - 10pm

Gabe's Family Restaurant
2044 Broadview Rd, (216) 741-4466
American, Cleveland
Eclipse Day Monday, 6am - 2pm

Petie's Family Restaurant
30150 Lakeshore Blvd Willowick, (440) 943-5900
American, Cleveland
Eclipse Day Monday, 7am - 3pm

Rowley Inn
1104 Rowley Ave, (216) 795-5345
American, Cleveland
Eclipse Day Monday, 7am - 10pm

Slyman's Restaurant and Deli
3106 St Clair Ave NE, (216) 621-3760
Deli, Cleveland
Eclipse Day Monday, 8am - 2:30pm

The Garden Family Restaurant
14957 Snow Rd Brook Park, (216) 267-5573
American, Cleveland
Eclipse Day Monday, 6am - 2pm

Tony's Family Restaurant
1515 W Pleasant Valley Rd Parma, (440) 842-2250
American, Cleveland
Eclipse Day Monday, 7am - 2pm

Victoria's Deli and Restaurant
6779 Ames Rd Parma, (440) 845-8922
American, Cleveland
Eclipse Day Monday, 8am - 2:30pm

Akron, OH
Totality Starts at: 3:14pm EDT
Totality Lasts for: 2:45
Totality Ends at: 3:17pm EDT
NWS Office: www.weather.gov/cle
City Website: www.akronohio.gov

Akron Family Restaurant
254 W Market St, (330) 376-0600
American, Akron
Eclipse Day Monday, 5:30am - 6pm

Arthur Treacher's Fish & Chips
1833 State Rd Cuyahoga Falls, (330) 923-8900
Fish & Chips, Akron
Eclipse Day Monday, 10:30am - 9:30pm

DeChecos Pizzeria
2075 S Main St, (330) 724-4835
Pizza, Akron
Eclipse Day Monday, 11am - 10pm

Diamond Deli
378 S Main St, (330) 762-5877
Deli, Akron
Eclipse Day Monday, 9am - 3pm

Ken Stewart's Grille
1970 W Market St, (330) 867-2555
Steak, Akron
Eclipse Day Monday, 11am - 10pm

Luigi's Restaurant
105 N Main St, (330) 253-2999
Italian, Akron
Eclipse Day Monday, 11am - 12am

Michael's AM
1562 Akron Peninsula Rd #120, (330) 929-3447
Breakfast, Akron
Eclipse Day Monday, 7:30am - 1:30pm

Steinly's Restaurant
235 E Waterloo Rd, (330) 773-6422
American, Akron
Eclipse Day Monday, 7am - 2pm

Swensons Drive-In
658 E Cuyahoga Falls Ave, (330) 928-8515
American, Akron
Eclipse Day Monday, 11am - 1am

Sushi Asia Gourmet
1375 N Portage Path, (234) 706-6750
Pan-Asian, Akron
Eclipse Day Monday, 11am - 8:30pm

The Blue Door Cafe & Bakery
1970 State Rd Cuyahoga Falls, (330) 926-9774
American, Akron
Eclipse Day Monday, 8am - 1:30pm

Uncle Tito's Mexican Grill
2215 E Waterloo Rd, (330) 208-0429
Mexican, Akron
Eclipse Day Monday, 11am - 10pm

Barberton, OH
Totality Starts at: 3:14pm EDT
Totality Lasts for: 2:41
Totality Ends at: 3:16pm EDT
NWS Office: www.weather.gov/cle
City: www.cityofbarberton.com

Al's Corner Restaurant
155 2nd St NW, (330) 475-7978
American, Barberton
Eclipse Day Monday, 11am - 2pm

Belgrade Gardens
401 E State St, (330) 745-0113
Chicken, Barberton
Eclipse Day Monday, 12–7pm

Green Diamond Grille & Pub
125 2nd St NW, (330) 745-1900
Steak, Barberton
Eclipse Day Monday, 11am - 9pm

Hopocan Gardens
4396 W Hopocan Ave Ext Norton, (330) 825-9923
Chicken, Barberton
Eclipse Day Monday, 11am - 9pm

Magic subs and gyros
540 Wooster Road North, (330) 805-4799
Sandwich, Barberton
Eclipse Day Monday, 10am - 8pm

The Coffee Pot
205 2nd St NW, (330) 745-2596
Breakfast, Barberton
Eclipse Day Monday, 6am - 12:30pm

Village Inn Chicken
4444 S Cleveland Massillon Rd, (330) 825-4553
Chicken, Norton
Eclipse Day Monday, 11am - 8pm

White House Chicken Systems Inc
180 Wooster Road North, (330) 745-0449
Chicken, Barberton
Eclipse Day Monday, 11am - 8pm

Painesville, OH
Totality Starts at: 3:14pm EDT
Totality Lasts for: 3:50
Totality Ends at: 3:18pm EDT
NWS Office: www.weather.gov/cle
City Website: www.painesville.com

Compadres Mexican Grill
1894 Mentor Ave, (440) 354-2265
Mexican, Painesville
Eclipse Day Monday, 11am - 9pm

El Taco Macho
1613 Mentor Ave, (440) 350-8226
Mexican, Painesville
Eclipse Day Monday, 11am - 10pm

Flavors Around The Square ~ F.A.T.S.
25 S St Clair St, (440) 350-3657
American, Painesville
Eclipse Day Monday, 7am - 3pm

Perry Family Restaurant
2736 N Ridge Rd, (440) 350-1960
American, Painesville
Eclipse Day Monday, 7am - 9pm

Redhawk Grille
7481 Auburn Rd #9703, (440) 354-4040
American, Painesville
Eclipse Day Monday, 11am - 12:30am

Sammy's Family Restaurant
625 River St, (440) 357-1417
American, Painesville
Eclipse Day Monday, 6am - 9pm

Scooter's World Famous Dawg House
9600 Blackbrook Rd Mentor, (440) 354-8480
Hot Dogs, Mentor, Opens in Spring
www.scootersworldfamousdawghouse.com

Never gaze at the sun without eye protection. Only remove eye protection during 100% totality.

The Sidewalk Cafe
1 S State St, (440) 352-0222
Breakfast, Painesville
Eclipse Day Monday, 6am - 3:30pm

Ashtabula, OH
Totality Starts at: 3:15pm EDT
Totality Lasts for: 3:47
Totality Ends at: 3:19pm EDT
NWS Office: www.weather.gov/cle
City: https://www.cityofashtabula.com

Aunt Judy's Diner
3475 Fargo Dr, (440) 536-5628
American, Ashtabula
Eclipse Day Monday, 6am - 2pm

Becker's Restaurant
1601 W Prospect Rd, (440) 993-1131
American, Ashtabula
Eclipse Day Monday, 7am - 9pm

Burrito Loco
2421 Lake Ave, (440) 964-2222
Tex-Mex, Ashtabula
Eclipse Day Monday, 10:45am - 7pm

Crows Nest
1257 Harmon Rd, (440) 964-2696
American, Ashtabula
Eclipse Day Monday, 10am - 9pm

Deb's Diner
3214 State Rd, (440) 997-0066
American, Ashtabula
Eclipse Day Monday, 6am - 2pm

Harbor Halcyon
1119 Bridge St, (440) 536-4291
American, Ashtabula
Eclipse Day Monday, 11:30am - 9pm

Harbor Perk Coffeehouse & Roasting Co.
1003 Bridge St, (440) 964-9277
Coffee Shop, Ashtabula
Eclipse Day Monday, 7am - 7pm

Lakeway Restaurant
729 Lake Ave, (440) 964-7176
Italian, Ashtabula
Eclipse Day Monday, 8am - 8pm

Mike's Gyros & More
2405 Lake Ave, (440) 536-5383
Greek, Ashtabula
Eclipse Day Monday, 10:30am - 8pm

The Little Pie Shop & Cafe
5050 Lake Rd W, (440) 536-4095
American, Ashtabula
Eclipse Day Monday, 6:30am - 3pm

Pennsylvania

Erie, PA
Totality Starts at: 3:16pm EDT
Totality Lasts for: 3:42
Totality Ends at: 3:20pm EDT
NWS Office: www.weather.gov/cle
City Website: www.visiterie.com

Avanti's
1662 W 8th St, (814) 456-3096
Breakfast, Erie
Eclipse Day Monday, 8am - 2pm

Bay House Oyster Bar & Restaurant
6 Sassafras Pier, (814) 413-7440
Oysters, Erie
Eclipse Day Monday, 11:30am - 10pm

Butch's Place Family Restaurant
3330 W 26th St, (814) 835-3372
American, Erie
Eclipse Day Monday, 6:45am - 2pm

Coney Island Lunch
3015 Buffalo Rd, (814) 899-0339
American, Erie
Eclipse Day Monday, 7am - 10pm

El Amigo Mexican Grill
333 State St, (814) 454-4600
Mexican, Erie
Eclipse Day Monday, 11am - 9:30pm

Habibi Mediterranean Cuisine
127 W 14th St, (814) 920-4756
Mediterranean, Erie
Eclipse Day Monday, 11am - 8pm

Lucky Louie's Beer Wieners
8238 Perry Hwy, (814) 314-9481
Wieners, Erie
Eclipse Day Monday, 11am - 8pm

McGarrey's Oakwood Cafe
1624 W 38th St, (814) 866-0552
American, Erie
Eclipse Day Monday, 11am - 10pm

New York Lunch Peninsula Drive
1525 Peninsula Dr, (814) 835-3647
American, Erie
Eclipse Day Monday, 11am - 7pm

Oliver's Rooftop
130 E Front St, (814) 920-9666
American, Erie
Eclipse Day Monday, 11am - 10pm

Porky's Pizzeria
302 W 8th St, (814) 454-5000
Pizza, Erie
Eclipse Day Monday, 11am - 2pm, 4pm–12am

Shirley's
5924 Old French Rd, (814) 315-1059
American, Erie
Eclipse Day Monday, 6am - 2pm

Shish Kabob
1202 French St, (814) 920-4442
Middle Eastern, Erie
Eclipse Day Monday, 11am - 11pm

Shoreline Bar and Grille
2 Sassafras Pier, (814) 636-1005
American, Erie
Eclipse Day Monday, 11:30am - 10pm

Teresa's Italian Deli
3203 Greengarden Blvd, (814) 864-5322
Deli, Erie
Eclipse Day Monday, 9am - 6pm

The Original Breakfast Place
2340 E 38th St, (814) 825-2727
Breakfast, Erie
Eclipse Day Monday, 6am - 1pm

Valerio's Italian Restaurant & Pizzeria
724 Powell Ave, (814) 833-8884
Pizza, Erie
Eclipse Day Monday, 11am - 10pm

Zodiac Dinor
2516 State St, (814) 455-3543
American, Erie
Eclipse Day Monday, 8am - 1pm

Never gaze at the sun without eye protection. Only remove eye protection during 100% totality.

Featured Parks and Attractions

The Cleveland Botanical Garden offers a variety of plants and flowers from around the world. The Garden has many different habitats, including a rainforest in the Eleanor Armstrong Smith Glasshouse, and the serene Japanese Garden with its tranquil pool and Shinto shrine. Younger visitors will enjoy the Hershey Children's Garden with fun and educational interactive exhibits. Everyone visiting on eclipse day will experience nearly four minutes of totality.

Cuyahoga Valley National Park, between Cleveland and Akron in Ohio, is brimming with great views and recreational opportunities. The park is known for its landscapes, featuring lush forests, rolling hills, and the winding Cuyahoga River. There are over 125 miles of hiking trails, including the Towpath Trail which follows the historic route of the Ohio & Erie Canal. Guests can also explore the park aboard the Cuyahoga Valley Scenic Railroad. This vast national park will be draped in the moon's shadow for 3:27 during the eclipse.

The Erie Zoo, in Erie, Pennsylvania, is an attraction for all ages. The zoo is home to over 400 animals, including giraffes, orangutans, and snow leopards. The Children's Zoo allows youngsters to interact with a variety of friendly creatures. In addition to the animal exhibits, the zoo also features a botanical garden and greenhouse. At the Erie Zoo, the total eclipse of the sun lasts for over three and a half minutes.

Malabar Farm State Park in Lucas, Ohio, was once the home of Pulitzer Prize-winning author Louis Bromfield. Visitors can tour the historic Bromfield home, explore the working farm, and enjoy scenic trails. The park offers camping facilities and the charming Pugh Cabin, famously known as the site of Humphrey Bogart and Lauren Bacall's wedding. At 3:12pm on eclipse day, the moon marries the sun and casts almost three minutes of shadow down on this Ohio state park.

In Oregon, Ohio, **Maumee Bay State Park** is a paradise for outdoor enthusiasts and nature lovers. The park features marshlands, meadows, and sandy beaches. Visitors can enjoy hiking on the park's nature trails, bird-watching, swimming, and playing golf on the 18-hole course. The park also offers a modern lodge with stunning lake views and well-equipped cottages and campgrounds. Totality starts at 3:12pm and lasts just over two minutes at this park.

The Rock & Roll Hall of Fame in Cleveland is a must-visit for music enthusiasts. This iconic museum celebrates the history and impact of rock & roll music and honors its most influential figures. Visitors can explore exhibits filled with memorabilia from many legendary artists. The museum's vast collection includes everything from Elvis Presley's jumpsuit to handwritten lyrics by The Beatles. Outside the museum, the total solar eclipse makes it's own music for nearly four minutes, don't miss it!

The parks and attractions section in this book are organized by cities, which are highlighted in bold. Under each city, you will find a curated list of locations presented in alphabetical order. These locations may be within the city or situated within a 30-50 mile radius, encompassing the surrounding region. This method of organization allows for easy navigation and planning, ensuring you can make the most of your solar eclipse experience.

Never gaze at the sun without eye protection. Only remove eye protection during 100% totality.

Parks and Attractions

Greenville, OH
Totality Starts at: 3:08pm EDT
Totality Lasts for: 3:56
Totality Ends at: 3:12pm EDT
NWS Office: www.weather.gov/iln
www.cityofgreenville.org

Bear's Mill
6450 Arcanum Bears Mill Rd
Greenville, OH 45331
www.bearsmill.org

Chenoweth Trails
440 Greenville-Nashville Rd
Greenville, OH 45331
www.mattlight72.com/about-chenoweth-trails/

Eldora Speedway
9726, 13929 OH-118, New Weston, OH 45348
Totality Starts at: 3:08pm EDT
Totality Lasts for: 3:58

Fort Saint Clair Park
135 Camden Rd, Eaton, OH 45320
Totality Starts at: 3:08pm EDT
Totality Lasts for: 3:24

Garst Museum
205 N Broadway St
Greenville, OH 45331
www.garstmuseum.org

Greenville City Park
108 Ave F
Greenville, OH 45331
www.cityofgreenville.org

Greenville Falls State Scenic River Area
9140 Covington-Gettysburg Rd
Covington, OH 45318
www.miamicountyparks.com

Lazy R Campground
8714 Old US Rte 36
Bradford, OH 45308
www.lazyrcampgrounds.com

Shawnee Prairie Preserve
4267 ST Rt 502
Greenville, OH 45331
www.darkecountyparks.org

Southside Splash Pad
Southside Park
Greenville, OH 45331
www.cityofgreenville.org

Stillwater Prairie Reserve
9750 OH-185, Covington, OH 45318
Totality Starts at: 3:08pm EDT
Totality Lasts for: 3:53

Dayton, OH
Totality Starts at: 3:09pm EDT
Totality Lasts for: 3:00
Totality Ends at: 3:12pm EDT
NWS Office: www.weather.gov/iln
City Website: www.daytonohio.gov

America's Packard Museum
420 S Ludlow St
Dayton, OH 45402
www.americaspackardmuseum.org

Aullwood Audubon Center and Farm
1000 Aullwood Rd, Dayton, OH 45414
Nature Sanctuary
www.aullwood.org

Boonshoft Museum of Discovery
2600 Deweese Pkwy
Dayton, OH 45414
www.boonshoftmuseum.org

Buck Creek State Park
1976 Buck Creek Ln, Springfield, OH 45502
Totality Starts at: 3:10pm EDT
Totality Lasts for: 2:34

Carillon Historical Park
1000 Carillon Blvd, Dayton, OH 45409
History Museum
www.daytonhistory.org

Carriage Hill MetroPark
7800 Shull Rd
Huber Heights, OH 45424
metroparks.org/places-to-go/carriage-hill/

Charleston Falls Preserve
2535 Ross Rd, Tipp City, OH 45371
Totality Starts at: 3:09pm EDT
Totality Lasts for: 3:13

Columbus Zoo and Aquarium
4850 W Powell Rd, Powell, OH 43065
Totality Starts at: 3:11pm EDT
Totality Lasts for: 2:00

Cox Arboretum MetroPark
6733 N Springboro Pike
Dayton, OH 45449
metroparks.org

Dayton Aviation National Historical Park
16 S Williams St
Dayton, OH 45402
www.nps.gov/daav/

Englewood MetroPark
4361 W National Rd
Dayton, OH 45414
metroparks.org/places-to-go/englewood/

Frankenstein's Castle
Patterson Blvd
Kettering, OH 45419
metroparks.org/places-to-go/hills-dales/

George Rogers Clark Park
930 S Tecumseh Rd
Springfield, OH 45506
www.clarkcountyparks.org

Glen Helen Nature Preserve
405 Corry St
Yellow Springs, OH 45387
www.glenhelen.org

Hartman Rock Garden
1905 Russell Ave, Springfield
Totality Starts at: 3:10pm EDT
Totality Lasts for: 2:34

Hueston Woods State Park
6301 Park Office Rd, College Corner
Totality Starts at: 3:08pm EDT
Totality Lasts for: 3:13

Island MetroPark
101 E Helena St
Dayton, OH 45405
metroparks.org

John Bryan State Park
3790 OH-370, Yellow Springs
Totality Starts at: 3:10pm EDT
Totality Lasts for: 2:04

Kiser Lake State Park
4889 OH-235, Conover, OH 45317
Totality Starts at: 3:09pm EDT
Totality Lasts for: 3:42

Miamisburg Mound Park
900 Mound Rd
Miamisburg, OH 45342
miamisburg-park.edan.io

National Museum of the US Air Force
1100 Spaatz St
Dayton, OH 45433
www.nationalmuseum.af.mil

Oakes Quarry Park
1267 E Xenia Dr
Fairborn, OH 45324
www.fairbornoh.gov

Possum Creek MetroPark
4790 Frytown Rd
Dayton, OH 45417
metroparks.org

Riverfront Park Miamisburg
3 North Miami Avenue
Miamisburg, OH 45342
www.playmiamisburg.com

RiverScape MetroPark
237 E Monument Ave
Dayton, OH 45402
metroparks.org

Scene75 Entertainment Center | Dayton
6196 Poe Ave
Dayton, OH 45414
www.scene75.com/dayton/

The Dayton Art Institute
456 Belmonte Park N
Dayton, OH 45405
www.daytonartinstitute.org

Never gaze at the sun without eye protection. Only remove eye protection during 100% totality.

Thomas A. Cloud Memorial Park
4707 Brandt Pike
Huber Heights, OH 45424
www.hhoh.org

Wegerzyn Gardens MetroPark
1301 E Siebenthaler Ave
Dayton, OH 45414
metroparks.org

Wright Brothers Memorial
2380 Memorial Rd
Dayton, OH 45424
www.nps.gov/daav/

Lima, OH
Totality Starts at: 3:09pm EDT
Totality Lasts for: 3:49
Totality Ends at: 3:13pm EDT
NWS Office: www.weather.gov/iwx
City Website: www.limaohio.com

Allen County Historical Society and Museum
620 W Market St
Lima, OH 45801
allencountymuseum.org

Armstrong Air & Space Museum
500 Apollo Dr
Wapakoneta, OH 45895
www.armstrongmuseum.org

Highest point in Ohio
2280 OH-540, Bellefontaine, OH 43311
Totality Starts at: 3:10pm EDT
Totality Lasts for: 3:43

Faurot Park
S Cole St
Lima, OH 45805
www.cityhall.lima.oh.us

Grand Lake St. Marys State Park
834 Edgewater Dr, St Marys, OH 45885
ohiodnr.gov
Totality Lasts for: 3:51

Huggy Bear Campground
9065 Ringwald Rd, Middle Point, OH 45863
www.huggybearcampground.com
Totality Lasts for: 3:17

Indian Lake State Park
13156 OH-235, Lakeview, OH 43331
Totality Starts at: 3:09pm EDT
Totality Lasts for: 3:56

Killdeer Plains Wildlife Area
19100 Co Hwy 115
Harpster, OH 43323
ohiodnr.gov

Lake Loramie State Park
4401 Fort Loramie-Swanders Rd, Minster, OH
ohiodnr.gov
Totality Lasts for: 3:58

Lima Bresler Reservoir
381-283 S Kemp Rd
Lima, OH 45806
Totality Lasts for: 3:45

The Bicycle Museum of America
7 W Monroe St, New Bremen, OH 45869
www.bicyclemuseum.com
Totality Lasts for: 3:57

Findlay, OH
Totality Starts at: 3:10pm EDT
Totality Lasts for: 3.42
Totality Ends at: 3:14pm EDT
NWS Office: www.weather.gov/cle
City Website: www.findlayohio.gov

Children's Museum of Findlay
1800 Tiffin Ave # 201
Findlay, OH 45840
www.cmfindlay.com

Emory Adams Park
1827 S Blanchard St
Findlay, OH 45840
www.findlayohio.gov

Fostoria Iron Triangle Railpark
Popular, 499 S Poplar St
Fostoria, OH 44830
fostoriairontriangle.com

Hancock Historical Museum
422 W Sandusky St
Findlay, OH 45840
hancockhistoricalmuseum.org

Independence Dam State Park
27722 County Rd 424, Defiance, OH 43512
ohiodnr.gov
Totality Lasts for: 1:59

Litzenberg Memorial Woods
6100 US-224
Findlay, OH 45840
hancockparks.com

Northwest Ohio Railroad Riverside Train
12505 County Rd 99
Findlay, OH 45840
www.nworrp.org

Oakwoods Nature Preserve
1400 Oakwood Avenue
Findlay, OH 45840
hancockparks.com

Riverbend Park
16618 Township Rd 208
Findlay, OH 45840
hancockparks.com

Rawson Park
720 River St
Findlay, OH 45840
www.findlayohio.gov

Riverside Park Pool
231 McManness Ave
Findlay, OH 45840
www.findlayohio.gov

Van Buren State Park
12259 Township Rd 218, Van Buren, OH 45889
Totality Starts at: 3:10pm EDT
Totality Lasts for: 3:33

Bowling Green, OH
Totality Starts at: 3:11pm EDT
Totality Lasts for: 3:00
Totality Ends at: 3:14pm EDT
NWS Office: www.weather.gov/cle
City Website: www.bgohio.org

Black Swamp Preserve
1014 S Maple St
Bowling Green, OH 43402
www.wcparks.org

Bowling Green City Park
520 Conneaut Ave
Bowling Green, OH 43402
www.bgohio.org

Carter Park
401 Campbell Hill Road
Bowling Green, OH 43402
www.bgohio.org

Carter Historic Farm
18331 Carter Rd
Bowling Green, OH 43402
www.wcparks.org

Fuller Preserve
12153 Cross Creek Rd
Bowling Green, OH 43402
www.woodcountyparkdistrict.org

Simpson Garden Park
1291 Conneaut Ave
Bowling Green, OH 43402
www.bgohio.org

White Star Quarry
901 S Main St, Gibsonburg, OH 43431
Totality Starts at: 3:11pm EDT
Totality Lasts for: 3:23

Wintergarden
615 S Wintergarden Rd
Bowling Green, OH 43402
www.bgohio.org/172/Wintergarden

Toledo, OH
Totality Starts at: 3:12pm EDT
Totality Lasts for: 2:00
Totality Ends at: 3:14pm EDT
NWS Office: www.weather.gov/cle
City Website: toledo.oh.gov

Buttonwood/Betty C. Black Recreation Area
27174 Hull Prairie Rd
Perrysburg, OH 43551
www.wcparks.org

Collins Park
755 York St
Toledo, OH 43605
toledo.oh.gov

Never gaze at the sun without eye protection. Only remove eye protection during 100% totality.

Fort Meigs Historic Site
29100 W River Rd
Perrysburg, OH 43551
www.fortmeigs.org

Imagination Station
1 Discovery Way
Toledo, OH 43604
www.imaginationstationtoledo.org

Independence Dam State Park
27722 County Rd 424, Defiance, OH 43512
Totality Starts at: 3:10pm EDT
Totality Lasts for: 1:55

International Park
Rails To Trails Next To Maumee River
Toledo, OH 43605
toledo.oh.gov

Jamie Farr Park
2140 N Summit St
Toledo, OH 43611
toledo.oh.gov

Mary Jane Thurston State Park
1466 State Rte 65, McClure, OH 43534
ohiodnr.gov
Totality Lasts for: 2:23

Maumee Bay State Park
1400 State Park Rd, Oregon, OH 43616
Totality Starts at: 3:12pm EDT
Totality Lasts for: 2:15

National Museum of the Great Lakes
1701 Front St
Toledo, OH 43605
www.nmgl.org

Navarre Park
1001 White St.
Toledo, OH 43605
toledo.oh.gov

Otsego Park
20000 W River Rd
Bowling Green, OH 43402
www.wcparks.org

Promenade Park
400 Water St
Toledo, OH 43604
toledo.oh.gov

Side Cut Metropark
1025 W River Rd
Maumee, OH 43537
metroparkstoledo.com

Toledo East / Stony Ridge KOA Journey
24787 Luckey Rd
Perrysburg, OH 43551
koa.com/campgrounds/toledo/

W. W. Knight Nature Preserve
29530 White Rd
Perrysburg, OH 43551
www.wcparks.org

Weirs Rapids
21095 Range Line Rd
Bowling Green, OH 43402
www.bgohio.org

Woodlands Park
429 E Boundary St
Perrysburg, OH 43551
ci.perrysburg.oh.us

Mansfield, OH
Totality Starts at: 3:12pm EDT
Totality Lasts for: 3:17
Totality Ends at: 3:15pm EDT
NWS Office: www.weather.gov/cle
City Website: ci.mansfield.oh.us

Alum Creek State Park Campground
2911 S Old State Rd, Delaware, OH 43015
Totality Starts at: 3:11pm EDT
Totality Lasts for: 2:08

Biblewalk
500 Tingley Ave
Mansfield, OH 44905
www.biblewalk.us

Bicentennial Park
Lex-Ontario Road, Lexington, OH 44904
Totality Starts at: 3:12pm EDT
Totality Lasts for: 3:12

Buckeye Imagination Museum
175 W 3rd St
Mansfield, OH 44903
buckeyeimaginationmuseum.org

Charles Mill Lake Park
1277A OH-430, Mansfield, OH 44903
Totality Starts at: 3:12pm EDT
Totality Lasts for: 3:08

Columbus Zoo and Aquarium
4850 W Powell Rd, Powell, OH 43065
Totality Lasts for: 3:11
Totality Ends at: 2:02pm EDT

Delaware State Park
5202 US-23, Delaware, OH 43015
ohiodnr.gov
Totality Lasts for: 3:00

Gorman Nature Center
2295 Lexington Ave
Mansfield, OH 44906
www.richlandcountyoh.gov

Kingwood Center Gardens
50 Trimble Rd
Mansfield, OH 44906
www.kingwoodcenter.org

Lowe-Volk Nature Center @ Crawford Park
2401 OH-598
Crestline, OH 44827
www.crawfordparkdistrict.org

Malabar Farm State Park
4050 Bromfield Rd, Lucas, OH 44843
Totality Starts at: 3:12pm EDT
Totality Lasts for: 2:46

Mohican State Park
3116 OH-3, Loudonville, OH 44842
Totality Starts at: 3:13pm EDT
Totality Lasts for: 2:15

Mt. Gilead State Park
4119 OH-95, Mt Gilead, OH 43338
ohiodnr.gov
Totality Lasts for: 3:05

North Lake Park
268 Hope Rd, Mansfield, OH 44903
Totality Starts at: 3:12pm EDT
Totality Lasts for: 3:19

Ohio Bird Sanctuary
3774 Orweiler Rd
Mansfield, OH 44903
www.ohiobirdsanctuary.com

The Mansfield Fire Museum
1265 W 4th St
Mansfield, OH 44903
themansfieldfiremuseum.com

The Ohio State Reformatory (Shawshank)
100 Reformatory Rd
Mansfield, OH 44905
www.mrps.org

Norwalk, OH
Totality Starts at: 3:12pm EDT
Totality Lasts for: 3:54
Totality Ends at: 3:16pm EDT
NWS Office: www.weather.gov/cle
City Website: www.norwalkoh.com

Augusta-Anne Olsen State Nature Preserve
4934 W River Rd
Wakeman, OH 44889
ohiodnr.gov

Cedar Point Amusement Park
1 Cedar Point Dr
Sandusky, OH 44870
www.cedarpoint.com

Castalia Quarry
8404 OH-101
Castalia, OH 44824
eriemetroparks.org

East Harbor State Park
1169 N Buck Rd, Lakeside Marblehead
Totality Starts at: 3:12pm EDT
Totality Lasts for: 3:29

Findley State Park
25381 OH-58, Wellington, OH 44090
Totality Starts at: 3:12pm EDT
Totality Lasts for: 3:44

Huron Lighthouse
Huron, OH 44839
Totality Starts at: 3:12pm EDT
Totality Lasts for: 3:49

Milan Wildlife Area
5096 Co Hwy 48
Norwalk, OH 44857
ohiodnr.gov

Never gaze at the sun without eye protection. Only remove eye protection during 100% totality.

Merry-Go-Round Museum
301 Jackson St
Sandusky, OH 44870
www.merrygoroundmuseum.org

Nickel Plate Beach
1 Nickel Plate Dr
Huron, OH 44839
Totality Lasts for: 3:50

Shoreline Park, llc
411 E Shoreline Dr
Sandusky, OH 44870
coastal.ohiodnr.gov/erie/shorelinepk

Thomas Edison Birthplace Museum
9 N Edison Dr
Milan, OH 44846
tomedison.org

Vermilion River Reservation: Mill Hollow
51211 N Ridge Rd
Vermilion, OH 44089
www.loraincountymetroparks.com

Avon, OH
Totality Starts at: 3:13pm EDT
Totality Lasts for: 3:53
Totality Ends at: 3:17pm EDT
NWS Office: www.weather.gov/cle
City Website: www.cityofavon.com

Black River Reservation Days Dam
2720 E 31st St
Lorain, OH 44055
www.loraincountymetroparks.com

Cascade Park
387 Furnace St
Elyria, OH 44035
www.loraincountymetroparks.com

Carlisle Reservation
12882 Diagonal Rd
Lagrange, OH 44050
www.loraincountymetroparks.com

French Creek Reservation
4540 French Creek Rd
Sheffield, OH 44054
www.loraincountymetroparks.com

Frostville Museum
24101 Cedar Point Rd
North Olmsted, OH 44070
www.olmstedhistoricalsociety.org

Huntington Reservation
28728 Wolf Picnic Area Dr
Bay Village, OH 44140
www.clevelandmetroparks.com

Lake Erie Nature & Science Center
28728 Wolf Rd
Bay Village, OH 44140
www.lensc.org

Lorain Harbor Lighthouse, llc
Lorain, OH 44052
lorainlighthouse.com
Totality Lasts for: 3:52

Miller Nature Preserve
2739 Center Rd
Avon, OH 44011
www.loraincountymetroparks.com

Miller Road Park
33760 Lake Rd
Avon Lake, OH 44012
www.avonlake.org

Rocky River Nature Center
24000 Valley Pkwy
North Olmsted, OH 44070
www.clevelandmetroparks.com

Sandy Ridge Reservation
6195 Otten Rd
North Ridgeville, OH 44039
www.loraincountymetroparks.com

Veterans' Memorial Park
32756 Lake Rd
Avon Lake, OH 44012
Totality Lasts for: 3:53

Cleveland, OH
Totality Starts at: 3:13pm EDT
Totality Lasts for: 3:50
Totality Ends at: 3:17pm EDT
NWS Office: www.weather.gov/cle
www.thisiscleveland.com

A Christmas Story House
3159 W 11th St, Cleveland, OH 44109
Museum
www.achristmasstoryhouse.com

Canal Exploration Center
7104 Canal Rd
Valley View, OH 44125
www.nps.gov

CanalWay Center
4524 E 49th St
Cuyahoga Heights, OH 44125
www.clevelandmetroparks.com

Cleveland Botanical Garden
11030 East Blvd
Cleveland, OH 44106
holdenfg.org

Cleveland Metroparks Zoo
3900 Wildlife Way
Cleveland, OH 44109
www.clevelandmetroparks.com/zoo

Cleveland Museum of Natural History
1 Wade Oval Dr, Cleveland, OH 44106
Natural history museum
www.cmnh.org

Cleveland Script Sign - Edgewater Park
1502 Abbey Ave
Cleveland, OH 44113
www.thisiscleveland.com

Cuyahoga Valley National Park
6947 Riverview Road, Peninsula, OH 44264
nps.gov/cuva/
Totality Lasts for: 3:27

Gordon Park
E. 72nd,
Cleveland, OH 44103
www.city.cleveland.oh.us

Great Lakes Science Center
601 Erieside Ave, Cleveland, OH 44114
Science Museum
greatscience.com

Greater Cleveland Aquarium
2000 Sycamore St
Cleveland, OH 44113
greaterclevelandaquarium.com

Headlands Beach State Park
9601 Headlands Rd, Mentor, OH 44060
Totality Starts at: 3:14pm EDT
Totality Lasts for: 3:51

Heinen's Downtown Cleveland
900 Euclid Ave
Cleveland, OH 44115
www.heinens.com/stores/downtown-cleveland/

Lake Metroparks Farmpark
8800 Euclid Chardon Rd
Kirtland, OH 44094
www.lakemetroparks.com/parks-trails/farmpark

Lake View Cemetery / The Haserot Angel
12316 Euclid Ave
Cleveland, OH 44106
www.lakeviewcemetery.com

Lakewood Park / Solstice Steps
14532 Lake Ave
Lakewood, OH 44107
www.onelakewood.com

Mosquito Lake State Park
1439 Wilson Sharpsville Rd, Cortland
ohiodnr.gov
Totality Lasts for: 2:22

NASA Glenn Visitor Center
601 Erieside Ave
Cleveland, OH 44114
www.greatscience.com

Nelson-Kennedy Ledges State Park
12440 OH-282, Garrettsville, OH 44231
Totality Starts at: 3:15pm EDT
Totality Lasts for: 2:55

North Chagrin Reservation
3037 Som Center Rd
Willoughby Hills, OH 44094
www.clevelandmetroparks.com

Never gaze at the sun without eye protection. Only remove eye protection during 100% totality.

Punderson State Park
11755 Kinsman Rd, Newbury Township
Totality Starts at: 3:14pm EDT
Totality Lasts for: 3:29

Rockefeller Park & Greenhouse
750 E 88th St
Cleveland, OH 44108
www.culturalgardens.org

Rock & Roll Hall of Fame
1100 E 9th St,
Cleveland, OH 44114
www.rockhall.com

Rocky River Park
20250 Beach Cliff Blvd
Rocky River, OH 44116
www.rrcity.com/city-parks

Squire's Castle
2844 River Rd
Willoughby Hills, OH 44094
www.clevelandmetroparks.com

The Cleveland Museum of Art
11150 East Blvd
Cleveland, OH 44106
www.clevelandart.org

Viaduct Park
Willis St
Bedford, OH 44146
www.clevelandmetroparks.com

Voinovich Bicentennial Park
800 E 9th Street Pier
Cleveland, OH 44114
www.northcoastharbor.org

Wendy Park
2800 Whiskey Island Dr
Cleveland, OH 44102
clevelandmetroparks.com

West Side Market
1979 W 25th St, Cleveland, OH 44113
Fresh Food Market
westsidemarket.org

Akron, OH
Totality Starts at: 3:14pm EDT
Totality Lasts for: 2:45
Totality Ends at: 3:17pm EDT
NWS Office: www.weather.gov/cle
City Website: www.akronohio.gov

Akron Art Museum
1 S High St
Akron, OH 44308
akronartmuseum.org

Akron Children's Museum
216 S Main St
Akron, OH 44308
akronkids.org

Akron Zoo
500 Edgewood Ave
Akron, OH 44307
www.akronzoo.org

Cascade Valley Metro Park-Chuckery Area
837 Cuyahoga St
Akron, OH 44313
www.summitmetroparks.org

F.A. Seiberling Nature Realm
1828 Smith Rd
Akron, OH 44313
www.summitmetroparks.org

Fort Island / Griffiths Park
461 Trunko Rd
Fairlawn, OH 44333
www.cityoffairlawn.com

Kent State University May 4 Visitors Center
300 Midway Dr
Kent, OH 44243
www.kent.edu/may4visitorscenter

Lake Milton State Park
16801 Mahoning Ave, Lake Milton, OH 44429
Totality Starts at: 3:15pm EDT
Totality Lasts for: 1:22

Plum Creek Park
590 Plum St
Kent, OH 44240
www.kentparksandrec.com

Portage Lakes State Park
5031 Manchester Rd, Akron, OH
ohiodnr.gov
Totality Lasts for: 2:24

Sand Run Metro Park
1300 Sand Run Rd
Akron, OH 44313
www.summitmetroparks.org

Stan Hywet Hall & Gardens
714 N Portage Path
Akron, OH 44303
www.stanhywet.org

Towner's Woods
2264 Ravenna Rd
Kent, OH 44240
www.co.portage.oh.us

West Branch State Park
5570 Esworthy Rd, Ravenna, OH 44266
Totality Starts at: 3:15pm EDT
Totality Lasts for: 2:19

Wingfoot Lake State Park
993 Goodyear Park Blvd, Mogadore, OH 44260
Totality Starts at: 3:14pm EDT
Totality Lasts for: 2:11

Barberton, OH
Totality Starts at: 3:14pm EDT
Totality Lasts for: 2:41
Totality Ends at: 3:16pm EDT
NWS Office: www.weather.gov/cle
www.cityofbarberton.com

Edgewood Park
1170 Liberty Ave
Barberton, OH 44203
www.cityofbarberton.com

Lake Anna Park
615 W Park Ave
Barberton, OH 44203
ww.cityofbarberton.com

Memorial Park
274 Grandview Ave
Wadsworth, OH 44281
www.wadsworthcity.com

Ohio Veterans' Memorial Park
8005 S Cleveland Massillon Rd
Clinton, OH 44216
www.ovmp.org

Silver Creek Metro Park
5000 Hametown Rd.
Norton, OH 44203
www.summitmetroparks.org

Tuscora Park
501 E Tuscarawas Ave
Barberton, OH 44203
www.cityofbarberton.com

Painesville, OH
Totality Starts at: 3:14pm EDT
Totality Lasts for: 3:50
Totality Ends at: 3:18pm EDT
NWS Office: www.weather.gov/cle
City Website: www.painesville.com

Beaty Landing
477 East Walnut Street, State Route 84
Painesville, OH 44077
www.lakemetroparks.com

Carol H Sweet Nature Center
5185 Corduroy Rd
Mentor, OH 44060
www.cmnh.org/Mentor-Marsh

Euclid Beach Park
16301 Lakeshore Blvd
Cleveland, OH 44110
www.euclidbeach.org

Fairport Harbor Lakefront Park
301 Huntington Beach Dr
Fairport Harbor, OH 44077
www.lakemetroparks.com

Fairport Marine Museum and Lighthouse
129 2nd St
Fairport Harbor, OH 44077
www.fairportharborlighthouse.org

Headlands Beach State Park
9601 Headlands Rd, Mentor, OH 44060
Totality Starts at: 3:14pm EDT
Totality Lasts for: 3:51

Never gaze at the sun without eye protection. Only remove eye protection during 100% totality.

Headland Dunes State Nature Preserve
9601 Headlands Rd
Mentor, OH 44060
naturepreserves.ohiodnr.gov/headlandsdunes

Helen Hazen Wyman Park
6101 Painesville Warren Rd
Concord, OH 44077
www.lakemetroparks.com

Kenneth J Sims Park
23131 Lakeshore Blvd
Euclid, OH 44123
www.cityofeuclid.com

Lake Erie Bluffs
2901 Clark Rd
Perry, OH 44081
www.lakemetroparks.com

Mentor Beach Park Pavilion
7779 Lakeshore Blvd
Mentor-On-The-Lake, OH 44060
cityofmentor.com

Osborne Park
38575 Lakeshore Blvd
Willoughby, OH 44094
willoughbyohio.com

Overlook Beach Park
Mentor-On-The-Lake, OH 44060
coastal.ohiodnr.gov/lake/overlookbeach
Totality Lasts for: 3:51

Paine Falls Park
5570 Paine Rd
Painesville, OH 44077
www.lakemetroparks.com

Painesville Township Park
1025 Hardy Rd
Painesville, OH 44077
www.lakemetroparks.com

Sunset Park
38327 Beachview Rd
Willoughby, OH 44094
Totality Lasts for: 3:51

Veterans Memorial Park
5730 Hopkins Rd
Mentor, OH 44060
www.lakemetroparks.com

Ashtabula, OH
Totality Starts at: 3:15pm EDT
Totality Lasts for: 3:47
Totality Ends at: 3:19pm EDT
NWS Office: www.weather.gov/cle
www.cityofashtabula.com

Arcola Creek Park
941 Dock Rd
Madison, OH 44057
www.lakemetroparks.com

Conneaut Township Park
480 Lake Rd
Conneaut, OH 44030
conneauttownshippark.com

Geneva State Park
4499 Padanarum Rd, Geneva, OH 44041
Totality Starts at: 3:14pm EDT
Totality Lasts for: 3:49

Geneva Township Park
5045 Lake Rd E
Geneva, OH 44041
genevatownshippark.org

Lake Shore Park
Lakeshore Dr
Ashtabula, OH 44004
www.lakeshoreparkashtabula.org

Pymatuning State Park (Ohio)
6100 Pymatuning Lake Rd, Andover, OH 44003
ohiodnr.gov
Totality Lasts for: 3:06

Saybrook Township Park
5941 Lake Rd W
Ashtabula, OH 44004
www.saybrookpark.org

Walnut Beach Park
West 1st Street &, Walnut Blvd
Ashtabula, OH 44004
coastal.ohiodnr.gov

Willow Lake Campground
3935 N Broadway
Geneva, OH 44041
www.willowlakecamping.com

Pennsylvania

Erie, PA
Totality Starts at: 3:16pm EDT
Totality Lasts for: 3:42
Totality Ends at: 3:20pm EDT
NWS Office: www.weather.gov/cle
City Website: www.visiterie.com

Asbury Woods
4105 Asbury Rd
Erie, PA 16506
www.asburywoods.org

Avonia Beach
Avonia Rd
Fairview, PA 16415
www.fairviewtownship.com

Bicentennial Tower Observation Deck
1 State St, Erie, PA 16507
(814) 455-6055
www.porterie.org/bicentennialtower/

Asbury Woods
4105 Asbury Rd, Erie, PA 16506
Nature Preserve
www.asburywoods.org

Erie Art Museum
20 E 5th St
Erie, PA 16507
erieartmuseum.org

Erie Bluffs State Park
11100 W Lake Rd, Lake City, PA 16423
dcnr.pa.gov/StateParks/
Totality Lasts for: 3:44

Erie KOA Holiday
6645 W Rd
McKean, PA 16426
koa.com/campgrounds/erie/

Erie Land Lighthouse
2 Lighthouse St
Erie, (814) 722-4610
www.presqueislelighthouse.org

Erie Maritime Museum
150 E Front St
Erie, PA 16507
eriemaritimemuseum.org

Erie Zoo
423 W 38th St,
Erie, PA (814) 864-4091
www.eriezoo.org

expERIEnce Children's Museum
420 French St
Erie, PA 16507
www.eriechildrensmuseum.org

Freeport Beach
1 Freeport Rd
North East, PA 16428
northeastborough.com

Headwaters Park
1927 Wager Rd
Erie, PA 16509
www.erieconservation.com

Lake Erie Arboretum at Frontier
1501 W 6th St
Erie, PA 16505
www.leaferie.org

Lake Erie Community Park
10192 W Lake Rd
Lake City, PA 16423
www.girardtownship.com

McClelland Park
2600 E 26th St
Erie, PA 16510
erietrails.org/mcclelland-park/

McKean Community Recreational Park
8798 Main St
McKean, PA 16426
www.mckeantownship.com

Pennsylvania Welcome Center
I-90, Tourist information center
North East, PA 16428
visitpa.com

Picnicana Park
9260 Old French Rd #9144
Waterford, PA 16441
summittownship.com

Never gaze at the sun without eye protection. Only remove eye protection during 100% totality.

Presque Isle Lighthouse
301 Peninsula Dr
Erie, PA 16505
www.presqueislelighthouse.org

Presque Isle Passage RV Park & Cabin Rentals
6300 Sterrettania Rd
Fairview, PA 16415
presqueislepassage.com

Presque Isle State Park
301 Peninsula Dr, Erie, PA 16505
Totality Starts at: 3:16pm EDT
Totality Lasts for: 3:45

Pymatuning State Park
2660 Williamsfield Rd, Jamestown
dcnr.pa.gov/StateParks/
Totality Lasts for: 3:37

Scott Park
2600 block West 6th
Erie, PA
erietrails.org/scott-park/

Shades Beach Park
7000 E Lake Rd
Erie, PA 16511
www.harborcreektownship.org

Six Mile Creek Park
7725 Clark Rd
Erie, PA 16510
erietrails.org/six-mile-creek-park/

Splash Lagoon
8091 Peach St
Erie, PA 16509
www.splashlagoon.com

Tom Ridge Environmental Center
301 Peninsula Dr
Erie, PA 16505
www.dcnr.pa.gov

Uncle John's Elk Creek Campground
575 Elk Creek Rd
Lake City, PA 16423
unclejohnselkcreekcamp.com

Watson-Curtze Mansion - Hagen History Center
356 W 6th St
Erie, PA 16507
www.eriehistory.org

Whitford Park
5400 Iroquois Ave
Erie, PA 16511
www.harborcreek.org

Wintergreen Gorge
3399 Cooper Rd
Erie, PA 16510
erietrails.org/wintergreen-gorge/

Woodcock Dam
22079 PA-198, Saegertown, PA 16433
Totality Starts at: 3:16pm EDT
Totality Lasts for: 2:44

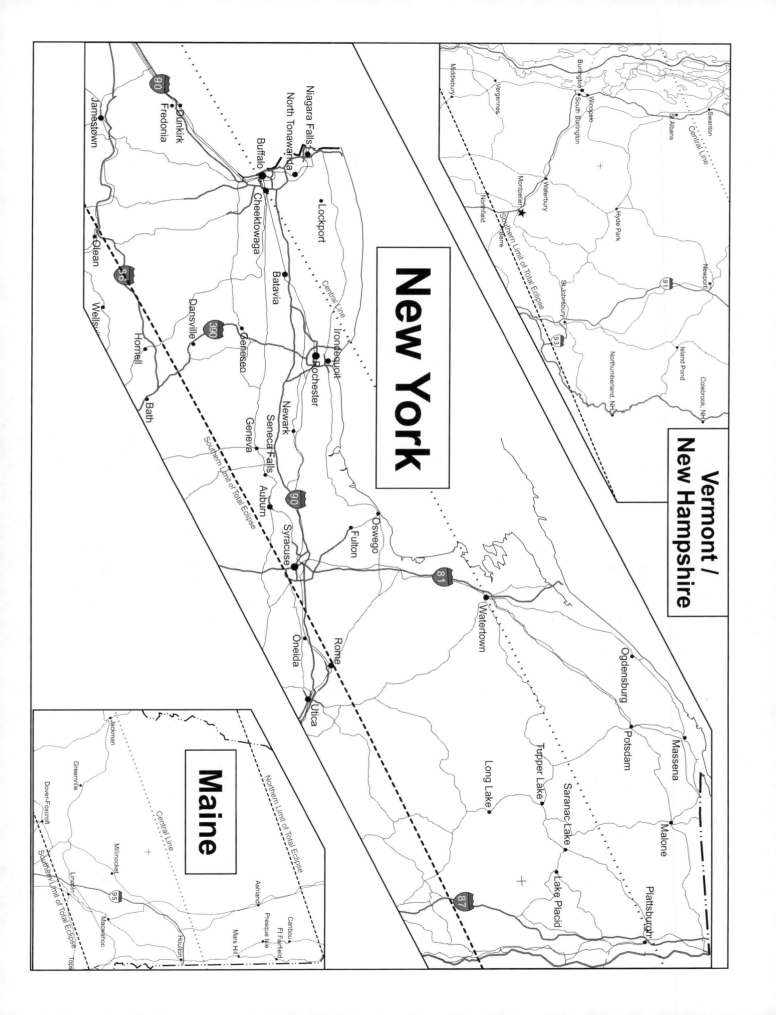

Featured Places to Eat

Alibaba Kebab in Buffalo, NY, features a fusion of Indian and Mediterranean. The menu has a selection of Alibaba Kebabs, including Boti Beef and Chicken Kebabs, Tandoori Chicken Tikka, and Bihari Kebab, all prepared with high-quality Halal meat and traditional spices.. Alibaba Kebab's Curry Dishes include flavorful options like Chicken Tikka Masala, Butter Chicken, and Palak Paneer (spinach). A 3:45 solar blackout awaits in Buffalo starting at 3:18pm; don't miss a moment by securing your meal in advance! Located at 900 William St, (716) 800-2222

Dogtown in Rochester, NY, is known for its specialty hot dogs, including the "Jindo" with kimchi, bacon, and sriracha mayo, and the "Chicago Bulldog" topped with mustard, relish, tomatoes, onions, green peppers, pepperoncini, and celery salt. Vegetarian customers can enjoy the "Veggie Wild Dog" with sweet Jamaica relish. Dogtown also serves tasty sandwiches, such as the Philly Cheese Steak. For a hearty meal check out the "Junkyard Cheeseburger Plate" with two cheeseburgers served over two sides. The cosmic ballet lasts for 3:40 in Rochester; so, order your food early and prepare for an unforgettable experience! Located at 691 Monroe Ave, (585) 271-6620

McSweeney's Red Hots in Plattsburgh, NY, focuses on burgers, hot dogs, and more. Customers can enjoy options like the classic Red Hot, which is a hot dog in a steamed bun with McS meat sauce, as well as a selection of hamburgers and cheeseburgers. They have hot sandwiches like the classic Steak Sandwich with peppers, onions, and cheese. Seafood lovers can opt for Fish on a Bun, Fish 'n' Chips, Clams, or Shrimp. Experience 3:34 of nature's grandest display in Plattsburgh, beginning at 3:25pm. Don't forget to order your food beforehand! Located at 7067 US-9, (518) 562-9309

Pho Capital in Montpelier, VT, features Vietnamese cuisine, with a focus on pho, a traditional Vietnamese noodle soup. The menu includes a variety of pho options, such as "Pho Tai Chin" with rare and well-done beef, and "Pho Tom" with shrimp. In addition to pho, the menu features a variety of rice and noodle dishes such as the "Combination Rice Plate" that includes grilled pork, shrimp, and egg rolls. Totality casts its spell for 1:42 in Montpelier. Order your food early and witness the solar phenomenon beginning at 3:27pm. Located at 107 State St, (802) 225-6183

Settler's Family Restaurant in Batavia, NY, has a diverse menu of American favorites, including salads, wraps, sandwiches, burgers, and dinner entrees. Salads include the Buffalo Chicken Salad and the Settler's Greek Souvlaki Salad. Wraps and pitas include the Chicken Cranberry Wrap with sliced almonds and dried cranberries. Dinner entrées feature dishes like Asian Grilled Salmon, Strip Steak, Philly Mac, and Homemade Mac & Cheese. Order your food and brace yourself for 3:43 of celestial magic in Batavia, starting at 3:19pm. Located at 353 W Main St, (585) 343-7443

The Polish Villa in Buffalo, NY, offers a delightful menu of authentic Polish cuisine. They have a selection of sandwiches like Polish Bologna Steak smothered in onions and dinners such as a Polish Platter with Golabki (cabbage rolls), Kielbasa (sausage), and pierogi, as well as specialties like Breaded Pork Chops and Marinated Chicken over Kluski noodles topped with mushroom gravy. The sun takes a 3:45 hiatus in Buffalo starting at 3:18pm; order your food early and enjoy the twilight phenomenon! Located at 2954 Union Rd, (716) 683-9460

Eclipse Track Notes - NY - VT - NH - ME

Mobility

In New York there are several interstates to aid in moving from one area to another. I-86 in the southwest corner runs east/west above Jamestown. I-90 runs east/west along Lake Erie and connects Erie, Pennsylvania with Buffalo, Rochester, and Syracuse. I-90 also provides a route into the path of totality from Albany, Schenectady, and Utica. I-81 connects Syracuse to Watertown. I-87 connects Plattsburgh to Albany in eastern New York. In Vermont, I-89 connects St. Albans City, Burlington and the capitol, Montpelier. I-91 in eastern Vermont is the north/south corridor. The eclipse crosses extreme northern New Hampshire where US-3 runs north/south along the border with Vermont. In Maine, I-95 provides a route into the total eclipse path and crosses US-1 near Houlton. US-1 runs north/south, connecting Caribou, Houlton and points south.

Jamestown, NY
Totality Starts at: 3:17pm EDT
Totality Lasts for: 2:52
Totality Ends at: 3:20pm EDT
NWS Office: www.weather.gov/buf
City Website: www.jamestownny.gov

AJ Texas Hots
824 Foote Ave, (716) 484-9646
Hot Dogs, Jamestown
Eclipse Day Monday, 10:30am - 10pm

Allen Street Diner
79 Allen St, (716) 484-4333
American, Jamestown
Eclipse Day Monday, 7am - 2:30pm

El Jarocho
323 Washington St, (716) 485-3085
Mexican, Jamestown
Eclipse Day Monday, 11am - 8pm

Jeremy's Belview
763 Foote Ave, (716) 664-5443
American, Jamestown
Eclipse Day Monday, 11am - 10pm

Lisciandro's Restaurant
207 N Main St, (716) 487-8925
American, Jamestown
Eclipse Day Monday, 7am - 1pm

Dunkirk, NY
Totality Starts at: 3:17pm EDT
Totality Lasts for: 3:43
Totality Ends at: 3:21pm EDT
NWS Office: www.weather.gov/buf
City Website: www.cityofdunkirk.com

Alma Latina
75 E 5th St, (716) 366-3030
Latin American, Dunkirk
Eclipse Day Monday, 10am - 7pm

Central Station Restaurant
332 Central Ave, (716) 366-8463
American, Dunkirk
Eclipse Day Monday, 7am - 3pm

Demetri's On the Lake
6-8 Lake Shore Dr W, (716) 366-4187
Greek, Dunkirk
Eclipse Day Monday, 8am - 8pm

El Azteca Restaurant
3953 Vineyard Dr, (716) 363-0300
Mexican, Dunkirk
Eclipse Day Monday, 11am - 3pm, 4–9:15pm

Jenna's 4th St Cafe
112 E 4th St, (716) 366-5360
American, Dunkirk
Eclipse Day Monday, 8am - 2pm

Lakeshore Grillworks Inc
436 Lake Shore Dr E, (716) 366-6600
American, Dunkirk
Eclipse Day Monday, 7:45am - 8pm

Mary's Deli
525 Main St, (716) 366-1445
Pizza, Dunkirk
Eclipse Day Monday, 11am - 9pm

Taqueria Mexicana
131 Central Ave, (716) 366-4030
Mexican, Dunkirk
Eclipse Day Monday, 11am - 8pm

Buffalo, NY
Totality Starts at: 3:18pm EDT
Totality Lasts for: 3:45
Totality Ends at: 3:22pm EDT
NWS Office: www.weather.gov/buf
City Website: www.buffalony.gov

99 Fast Food Restaurant
3398 Bailey Ave, (716) 836-6058
Vietnamese, Buffalo
Eclipse Day Monday, 10:30am - 9pm

Alibaba Kebab
900 William St, (716) 800-2222
Halal, Buffalo
Eclipse Day Monday, 11am - 12am

Allen Burger Venture
175 Allen St, (716) 768-0386
American, Buffalo
Eclipse Day Monday, 11:30am - 10pm

Never gaze at the sun without eye protection. Only remove eye protection during 100% totality.

Anchor Bar
1047 Main St, (716) 883-1134
Original Buffalo Wing, Buffalo
Eclipse Day Monday, 11am - 8:30pm

Antoinette's Sweets Inc.
5981 Transit Rd Depew, (716) 684-2376
Candy Store, Buffalo
Eclipse Day Monday, 11am - 9pm

Bada Bing
42 W Chippewa St, (716) 853-2464
American, Buffalo
Eclipse Day Monday, 11am - 11pm

Kostas Family Restaurant
1561 Hertel Ave, (716) 838-5225
Greek & American, Buffalo
Eclipse Day Monday, 7am - 9pm

Louie's Texas Red Hots
777 Harlem Rd West Seneca, (716) 823-7779
Hot Dogs, Buffalo
Eclipse Day Monday, 7am - 9pm

Mulberry Italian Ristorante
64 Jackson Ave Lackawanna, (716) 822-4292
Italian, Buffalo
Eclipse Day Monday, 11:30am - 3pm, 4–9pm

Mythos Restaurant
510 Elmwood Ave, (716) 886-9175
Greek, Buffalo
Eclipse Day Monday, 8am - 9pm

Nick's Place
504 Amherst St, (716) 871-1772
Breakfast, Buffalo
Eclipse Day Monday, 7am - 2pm

Polish Villa
2954 Union Rd, (716) 683-9460
Polish, Buffalo
Eclipse Day Monday, 8am - 8pm

Sophia's Restaurant
749 Military Rd, (716) 447-9661
Breakfast, Buffalo
Eclipse Day Monday, 7am - 3pm

Sun Cuisines
1989 Niagara St, (716) 447-0202
Burmese, Buffalo
Eclipse Day Monday, 11:30am - 9pm

Swan Street Diner
700 Swan St, (716) 768-1823
American, Buffalo
Eclipse Day Monday, 7am - 3pm

Ted's Hot Dogs
7018 Transit Rd, (716) 633-1700
Hot Dogs, Buffalo
Eclipse Day Monday, 11am - 8pm

Teton Kitchen Thai & Japanese Cuisine
415 Dick Rd Depew, (716) 393-3720
Sushi, Buffalo
Eclipse Day Monday, 11am - 10pm

The Place
229 Lexington Ave, (716) 882-7522
American, Buffalo
Eclipse Day Monday, 11:30am - 11pm

Niagara Falls, NY
Totality Starts at: 3:18pm EDT
Totality Lasts for: 3:28
Totality Ends at: 3:21pm EDT
NWS Office: www.weather.gov/buf
City: www.niagarafallsusa.com

El Cubilete
2050 Cayuga Drive Extension, (716) 297-4500
Mexican, Niagara Falls
Eclipse Day Monday, 11am - 8pm

Hyde Park Cafe
311 Hyde Park Blvd, (716) 297-7687
American, Niagara Falls
Eclipse Day Monday, 7:30am - 2:30pm

Indian Kitchen King
610 Main St, (716) 215-6994
Indian, Niagara Falls
Eclipse Day Monday, 11am - 8:30pm

La Galera Mexican Restaurant
8215 Niagara Falls Blvd, (716) 283-0005
Mexican, Niagara Falls
Eclipse Day Monday, 11:30am - 9pm

Mighty Taco
2591 Military Rd, (716) 297-7198
Mexican, Niagara Falls
Eclipse Day Monday, 10am - 11pm

Mister B's Restaurant
2201 Hyde Park Blvd, (716) 298-4028
Pizza, Niagara Falls
Eclipse Day Monday, 11am - 11pm

Mom's Family Restaurant
2410 Military Rd #1563, (716) 283-6667
American, Niagara Falls
Eclipse Day Monday, 7am - 3pm

Nelsons Cafe
2298 River Rd, Niagara Falls, NY 14304
American, Niagara Falls
Eclipse Day Monday, 7am - 2pm

Niagara Falls Buffet
7325 Niagara Falls Blvd, (716) 283-1318
Buffet, Niagara Falls
Eclipse Day Monday, 11am - 9pm

Sammy's Pizzeria
1400 Hyde Park Blvd, (716) 297-8442
Pizza, Niagara Falls
Eclipse Day Monday, 11am - 2am

Sunshine Cafe
8649 Buffalo Ave, (716) 236-0051
Breakfast, Niagara Falls
Eclipse Day Monday, 6am - 2pm

The Griffon Gastropub
2470 Military Rd, (716) 236-7474
Gastropub, Niagara Falls
Eclipse Day Monday, 11am - 2am

The Why Coffee Shop
1317 Main St, (716) 423-3145
Breakfast, Niagara Falls
Eclipse Day Monday, 8am - 1:30pm

Batavia, NY
Totality Starts at: 3:19pm EDT
Totality Lasts for: 3:43
Totality Ends at: 3:23pm EDT
NWS Office: www.weather.gov/buf
City: www.batavianewyork.com

Bourbon & Burger Co.
9 Jackson St, (585) 219-4242
American, Batavia
Eclipse Day Monday, 11am - 9pm

Cinquino's Pizza
314 Ellicott St, (585) 343-2447
Pizza, Batavia
Eclipse Day Monday, 11am - 9pm

Los Compadres Mexican Taqueria
40 Oak St, (585) 250-4067
Mexican, Batavia
Eclipse Day Monday, 11am - 9pm

Miss Batavia Diner
566 E Main St, (585) 343-9786
American, Batavia
Eclipse Day Monday, 6am - 2pm

Pizza 151
8351 Lewiston Rd, (585) 344-2400
Pizza, Batavia
Eclipse Day Monday, 10am - 10pm

Pok-A-Dot
229 Ellicott St, (585) 343-6775
American, Batavia
Eclipse Day Monday, 9am - 8pm

Rancho Viejo
12 Ellicott St, (585) 343-3903
Mexican, Batavia
Eclipse Day Monday, 11am - 9pm

Settler's Family Restaurant
353 W Main St, (585) 343-7443
American, Batavia
Eclipse Day Monday, 6am - 8pm

Town & Country Restaurant
5025 E Main Street Rd, (585) 343-3304
American, Batavia
Eclipse Day Monday, 6am - 8pm

Yume Asian Bistro
4140 Veterans Memorial Dr, (585) 345-9863
Asian, Batavia
Eclipse Day Monday, 11am - 9:30pm

Never gaze at the sun without eye protection. Only remove eye protection during 100% totality.

Rochester, NY
Totality Starts at: 3:20pm EDT
Totality Lasts for: 3:40
Totality Ends at: 3:23pm EDT
NWS Office: www.weather.gov/buf
City Website: www.visitrochester.com

Aladdin's Natural Eatery
646 Monroe Ave, (585) 442-5000
Mediterranean, Rochester
Eclipse Day Monday, 11am - 9pm

Bella Pasta Restaurant and Catering
2500 Ridgeway Ave, (585) 340-6100
Italian, Rochester
Eclipse Day Monday, 11am - 9pm

Charlie Riedel's Restaurant
1843 Empire Blvd Webster, (585) 671-4320
American, Webster
Eclipse Day Monday, 11am - 9pm

CRISP Rochester
819 S Clinton Ave, (585) 978-7237
American, Rochester
Eclipse Day Monday, 11am - 9pm

Dinosaur Bar-B-Que
99 Court St, (585) 325-7090
BBQ, Rochester
Eclipse Day Monday, 11am - 9pm

Dogtown
691 Monroe Ave, (585) 271-6620
German-style franks, Rochester
Eclipse Day Monday, 11am - 10:45pm

East Ridge Family Restaurant
1925 East Ridge Road, (585) 338-7900
American, Rochester
Eclipse Day Monday, 7am - 8pm

El Latino Restaurant
1020 Chili Ave, (585) 235-3110
Dominican, Rochester
Eclipse Day Monday, 10am - 8pm

Highland Park Diner
960 S Clinton Ave, (585) 461-5040
American, Rochester
Eclipse Day Monday, 8am - 3pm

Jim's On Main
785 E Main St, (585) 442-4172
American, Rochester
Eclipse Day Monday, 5am - 3pm

Jines Restaurant
658 Park Ave, (585) 461-1280
American, Rochester
Eclipse Day Monday, 7am - 3pm

John's Tex Mex
426 South Ave, (585) 232-5830
Tex-Mex, Rochester
Eclipse Day Monday, 11:30am - 10pm

Lakeside Haven Family Restaurant
3212 Lake Ave, (585) 663-8570
American, Rochester
Eclipse Day Monday, 6:30am - 2pm

Mi Viejo San Juan Restaurant
1143 Joseph Ave, (585) 467-1205
Puerto Rican, Rochester
Eclipse Day Monday, 10am - 6pm

Olympia Family Restaurant
1100 Flynn Rd, (585) 663-3071
American, Rochester
Eclipse Day Monday, 8am - 8pm

Parkway Family Restaurant
697 Ling Rd, (585) 663-9689
American, Rochester
Eclipse Day Monday, 7am - 2pm

Peppermill Restaurant
1776 Dewey Ave, (585) 621-4527
American, Rochester
Eclipse Day Monday, 7am - 2pm

Pita Restaurant
1378 Mt Hope Ave, (585) 271-7482
Lebanese, Rochester
Eclipse Day Monday, 11am - 8pm

Pudgie's Pizzeria
1753 N Goodman St, (585) 266-6605
Pizza, Rochester
Eclipse Day Monday, 11am - 9pm

Ricci's Family Restaurant
3166 Latta Rd, (585) 227-6750
Italian, Rochester
Eclipse Day Monday, 11am - 8:30pm

Roam Cafe
260 Park Ave, (585) 360-4165
American, Rochester
Eclipse Day Monday, 11:30am - 11pm

SEA Restaurant
1675 Mt Hope Ave, (585) 461-4154
Vietnamese, Rochester
Eclipse Day Monday, 11am - 9pm

The Original Charbroil House Restaurant
1395 Island Cottage Rd, (585) 663-3860
American, Rochester
Eclipse Day Monday, 11:30am - 8pm

Oswego, NY
Totality Starts at: 3:21pm EDT
Totality Lasts for: 3:30
Totality Ends at: 3:25pm EDT
NWS Office: www.weather.gov/buf
City Website: townofoswego.com

Azteca Mexican Grill
53 E Bridge St, (315) 341-7045
Mexican, Oswego
Eclipse Day Monday, 11:30am - 9pm

Canale's Restaurant
156 W Utica St, (315) 343-3540
Italian, Oswego
Eclipse Day Monday, 12–8pm

Fajita Grill
244 W Seneca St, (315) 326-0224
Southwestern, Oswego
Eclipse Day Monday, 11am - 9pm

Kiyomi
311 W Seneca St, (315) 343-8889
Japanese, Oswego
Eclipse Day Monday, 11am - 3pm, 4:30–9:30pm

Maria's Family Restaurant
111 W 2nd St, (315) 216-4562
American, Oswego
Eclipse Day Monday, 7am - 2pm

Oswego Sub Shop
106 W Bridge St, (315) 343-1233
American, Oswego
Eclipse Day Monday, 10:30am - 12am

Rudy's Lakeside Drive-In
78 Co Rte 89, (315) 343-2671
Seafood, rudyshot.com, Oswego
Seasonal, opens Mid-March

The Red Sun Fire Roasting Co
207 W 1st St, (315) 343-2418
American, Oswego
Eclipse Day Monday, 11:30am - 2pm, 4:30–8pm

Wade's Diner
176 E 9th St, (315) 343-6429
Breakfast, Oswego
Eclipse Day Monday, 6am - 12pm

Syracuse, NY
Totality Starts at: 3:22pm EDT
Totality Lasts for: 1:45
Totality Ends at: 3:24pm EDT
NWS Office: www.weather.gov/bgm
City Website: www.visitsyracuse.com

Bull and Bear Roadhouse
6402 Collamer Rd East Syracuse, (315) 437-2855
American, Syracuse
Eclipse Day Monday, 11:30am - 10pm

Don Juan Cafe Restaurant
102 Grand Ave, (315) 472-1770
Puerto Rican, Syracuse
Eclipse Day Monday, 11am - 7pm

Funk 'n Waffles
307-13 S Clinton St, (315) 474-1060
Chicken & Waffles, Syracuse
Eclipse Day Monday, 9am - 12am

Guadalajara Mexican restaurant
324 W Water St, (315) 552-1300
Mexican, Syracuse
Eclipse Day Monday, 11am - 9:30pm

Heid's of Liverpool
305 Oswego St Liverpool, (315) 451-0786
Hot Dogs & American, Syracuse
Eclipse Day Monday, 10am - 9pm

Joey's Italian Restaurant
6594 Thompson Rd, (315) 432-0315
Italian, Syracuse
Eclipse Day Monday, 11am - 9pm

Never gaze at the sun without eye protection. Only remove eye protection during 100% totality.

Mi Casita Restaurant
1614 Lodi St, (315) 870-3392
Puerto Rican, Syracuse
Eclipse Day Monday, 9am - 8pm

Nestico's Too
4105 W Genesee St, (315) 487-5864
American, Syracuse
Eclipse Day Monday, 7am - 2pm

Pastabilities
311 S Franklin St, (315) 474-1153
Italian, Syracuse
Eclipse Day Monday, 11am - 2pm, 4–9pm

Phoebe's Restaurant & Coffee Lounge
900 E Genesee St, (315) 475-5154
American, Syracuse
Eclipse Day Monday, 9am - 2:30pm

Red Chili Restaurant - Syracuse
2740 Erie Blvd E, (315) 446-2882
Chinese, Syracuse
Eclipse Day Monday, 11am - 9:30pm

Rise N Shine Diner Westcott
500 Westcott St, (315) 907-3710
American, Syracuse
Eclipse Day Monday, 7am - 3pm

Stella's Diner
110 Wolf St, (315) 425-0353
American, Syracuse
Eclipse Day Monday, 6am - 2:30pm

The Spinning Wheel Restaurant
7384 Thompson Rd North Syracuse, (315) 458-3222
American, Syracuse
Eclipse Day Monday, 11am - 11pm

Watertown, NY
Totality Starts at: 3:22pm EDT
Totality Lasts for: 3:38
Totality Ends at: 3:26pm EDT
NWS Office: www.weather.gov/buf
City Website: www.watertown-ny.gov

Cam's Pizzeria
25 Public Square, (315) 779-8900
Pizza, Watertown
Eclipse Day Monday, 11am - 9pm

Friede's Restaurant
455 Court St, (315) 221-4270
American, Watertown
Eclipse Day Monday, 6am - 2pm

Hacienda Authentic Mexican Restaurant
821 Arsenal St, (315) 608-3002
Mexican, Watertown
Eclipse Day Monday, 11am - 9:30pm

Jean's Beans
259 Eastern Blvd, (315) 788-7460
American, Watertown
Eclipse Day Monday, 9am - 7pm

Lotus Restaurant
1283 Arsenal St, (315) 788-3888
Vietnamese, Watertown
Eclipse Day Monday, 11am - 8pm

Mo's Place
345 Factory St, (315) 782-5503
American, Watertown
Eclipse Day Monday, Open 24 hours

Shorty's Place
1280 Coffeen St, (315) 782-7878
American, Watertown
Eclipse Day Monday, 7am - 2pm

Vito's Gourmet
3 Public Square, (315) 779-8486
Sandwich, Watertown
Eclipse Day Monday, 10:30am - 3pm

Plattsburgh, NY
Totality Starts at: 3:25pm EDT
Totality Lasts for: 3:34
Totality Ends at: 3:29pm EDT
NWS Office: www.weather.gov/bvt
City: www.cityofplattsburgh-ny.gov

Aleka's
103 Margaret St, (518) 310-3200
Greek, Plattsburgh
Eclipse Day Monday, 11am - 9pm

Anthony's Restaurant & Bistro
538 State Rte 3, (518) 561-6420
American, Plattsburgh
Eclipse Day Monday, 11:30am - 2pm, 4:30–8pm

Bazzano's Pizza
5041 S Catherine St, (518) 562-8586
Pizza, Plattsburgh
Eclipse Day Monday, 10am - 9pm

Clare & Carl's Hot Dog Stand
4729 US-9, (518) 561-1163
Hot Dogs, Plattsburgh
Seasonal, Opens in Spring

Duke's Diner
8 Tom Miller Rd, (518) 563-5134
American, Plattsburgh
Eclipse Day Monday, 7am - 2pm

Gus' Red Hots
5 Commodore MacDonough Hwy, (518) 735-0936
American, Plattsburgh
Eclipse Day Monday, 7am - 7pm

Hungry Bear Restaurant
2 Big Hank Plaza Rd, (518) 562-9144
American, Plattsburgh
Eclipse Day Monday, 7am - 3pm

Mainely Lobster & Seafood
1785 Military Turnpike # 1, (518) 562-7837
Seafood, Plattsburgh
Eclipse Day Monday, 11am - 8pm

McSweeney's Red Hots & Restaurant
7067 US-9, (518) 562-9309
American, Hot Dogs, Plattsburgh
Eclipse Day Monday, 11am - 8pm

Michigans Plus
313 Cornelia St, (518) 561-0537
American, Plattsburgh
Eclipse Day Monday, 7am - 3pm

Mickey's Restaurant & Lounge
26 Riley Ave, (518) 561-0066
American, Plattsburgh
Eclipse Day Monday, 11am - 9pm

Penny's Homestyle Cooking
364 Tom Miller Rd, (518) 310-3047
American, Plattsburgh
Eclipse Day Monday, 5am - 2pm

The Pepper
13 City Hall Pl, (518) 566-4688
Mexican, Plattsburgh
Eclipse Day Monday, 11:30am - 8pm

Vermont

Burlington, VT
Totality Starts at: 3:26pm EDT
Totality Lasts for: 3:13
Totality Ends at: 3:29pm EDT
NWS Office: www.weather.gov/bvt
City Website: www.burlingtonvt.gov

Al's French Frys
1251 Williston Rd, (802) 862-9203
American, S Burlington
Eclipse Day Monday, 10:30am - 11pm

American Flatbread Burlington Hearth
115 St Paul St, (802) 861-2999
Pizza, Burlington
Eclipse Day Monday, 11:30am - 10pm

Burlington Bagel Bakery
93 Church St, (802) 497-1530
American, Burlington
Eclipse Day Monday, 8am - 2:30pm

Masala Elaichi Indian Restaurant & Bar
207 Colchester Ave, (802) 540-2064
Indian, Burlington
Eclipse Day Monday, 10:30am - 2:30, 4:30–9:30

Myer's Bagel Bakery
377 Pine St, (802) 863-5013
Bagels, Burlington
Eclipse Day Monday, 6am - 2pm

Parkway Diner
1696 Williston Rd, (802) 540-9222
American, S Burlington
Eclipse Day Monday, 7am - 2pm

Saigon Kitchen
112 North St, (802) 540-0417
Vietnamese, Burlington
Eclipse Day Monday, 11:30am - 9pm

Never gaze at the sun without eye protection. Only remove eye protection during 100% totality.

Sherpa Kitchen
119 College St, (802) 881-0550
Nepalese, Burlington
Eclipse Day Monday, 11am - 2pm, 5–9pm

Shanty On The Shore
181 Battery St, (802) 864-0238
Seafood, Burlington
Eclipse Day Monday, 11:30am - 8:30pm

Speeder & Earl's Coffee
412 Pine St, (802) 658-6016
Coffee Shop, Burlington
Eclipse Day Monday, 6:30am - 6pm

The Farmhouse Tap & Grill
160 Bank St, (802) 859-0888
American, Burlington
Eclipse Day Monday, 11am - 10pm

The Firebird Cafe
1 Main St Essex Junction, (802) 316-4265
American, Essex Junction
Eclipse Day Monday, 8am - 3pm

The Skinny Pancake Burlington
60 Lake St, (802) 540-0188
Breakfast, Burlington
Eclipse Day Monday, 8am - 8pm

The Spot
210 Shelburne Rd, (802) 540-1778
Breakfast, Burlington
Eclipse Day Monday, 8am - 3pm

St Albans City, VT
Totality Starts at: 3:26pm EDT
Totality Lasts for: 3:33
Totality Ends at: 3:29pm EDT
NWS Office: www.weather.gov/bvt
City Website: www.stalbansvt.com

Lucky Buffet
101 Lake St, (802) 527-8388
Chinese, St Albans City
Eclipse Day Monday, 11am - 9pm

Maple City Diner
17 Swanton Rd, (802) 528-8400
American, St Albans City
Eclipse Day Monday, 7am - 3pm

Mimmo's Pizzeria & Restaurant - St. Albans
22 S Main St, (802) 524-2244
Italian, St Albans City
Eclipse Day Monday, 11am - 8pm

Pie In the Sky
267 Swanton Rd, (802) 524-5442
Pizza, St Albans City
Eclipse Day Monday, 11am - 8pm

St Albans Diner
8 Swanton Rd, (802) 528-8505
American, St Albans City
Eclipse Day Monday, 7am - 8pm

Thai House Restaurant
333 Swanton Rd, (802) 524-0999
Thai, St Albans City
Eclipse Day Monday, 11am - 2:30pm, 4–9pm

The Traveled Cup
94 N Main St, (802) 524-2037
Coffee Shop, St Albans City
Eclipse Day Monday, 7:30am - 5:30pm

Montpelier, VT
Totality Starts at: 3:27pm EDT
Totality Lasts for: 1:42
Totality Ends at: 3:29pm EDT
NWS Office: www.weather.gov/bvt
City Website: www.montpelier-vt.org

Buddy's Famous
15 Barre St, (802) 225-6400
American, Montpelier
Eclipse Day Monday, 11am - 8pm

The Mad Taco
72 Main St, (802) 225-6038
Mexican, Montpelier
Eclipse Day Monday, 12–8pm

Pho Capital
107 State St, (802) 225-6183
Vietnamese, Montpelier
Eclipse Day Monday, 11am - 9pm

Positive Pie Inc.
22 State St, (802) 229-0453
Pizza, Montpelier
Eclipse Day Monday, 11am - 8:30pm

Sarducci's
3 Main St, (802) 223-0229
Italian, Montpelier
Eclipse Day Monday, 11:30am - 9pm

The Skinny Pancake Montpelier
89 Main St, (802) 262-2253
Breakfast, Montpelier
Eclipse Day Monday, 8am - 8pm

Wayside Restaurant Bakery & Creamery
1873 US-302, (802) 223-6611
American, Montpelier
Eclipse Day Monday, 7am - 8:30pm

Northumberland, NH
Totality Starts at: 3:28pm EDT
Totality Lasts for: 1:49
Totality Ends at: 3:30pm EDT
NWS Office: www.weather.gov/gyx
City: www.northumberlandnh.org

North Country Family Restaurant
12 Main St, Groveton, (603) 636-1511
American, Groveton, Northumberland
Eclipse Day Monday, 7am - 2pm

Maine

New Hampshire

Caribou, ME
Totality Starts at: 3:32pm EDT
Totality Lasts for: 2:11
Totality Ends at: 3:34pm EDT
NWS Office: www.weather.gov/car
City: www.cariboumaine.org

Burger Boy
234 Sweden St, (207) 498-2329
American, Caribou
Eclipse Day Monday, 10:30am - 7pm

Colebrook, NH
Totality Starts at: 3:28pm EDT
Totality Lasts for: 2:59
Totality Ends at: 3:31pm EDT
NWS Office: www.weather.gov/gyx
City Website: colebrooknh.org

Black Bear Tavern
151 Main St, (603) 237-5521
American, Colebrook
Eclipse Day Monday, 11:30am - 9pm

House of Pizza
1 Parsons St, (603) 237-5256
Pizza, Colebrook
Eclipse Day Monday, 11am - 9pm

Caribou Bowl - A - Drome
97 Bennett Dr, (207) 498-3386
American, Caribou
Eclipse Day Monday, 11am - 12am

Cindy's Sub Shop
264 Sweden St, (207) 498-6021
Sandwich, Caribou
Eclipse Day Monday, 7:30am - 7pm

Jade Palace Restaurant
30 Skyway Dr, (207) 498-3648
Chinese, Caribou
Eclipse Day Monday, 11am - 9pm

Never gaze at the sun without eye protection. Only remove eye protection during 100% totality.

Houlton, ME
Totality Starts at: 3:32pm EDT
Totality Lasts for: 3:20
Totality Ends at: 3:35pm EDT
NWS Office: www.weather.gov/car
City: www.houlton-maine.com

Elm Tree Diner
146 Bangor St, (207) 254-2209
American, Houlton
Eclipse Day Monday, 5:30am - 8pm

Houlton Big Stop Restaurant
267 North St, (207) 521-0290
American, Houlton
Eclipse Day Monday, 7am - 8pm

Tang's Chinese Cuisine
60 North St, (207) 532-9981
Chinese, Houlton
Eclipse Day Monday, 11am - 8pm

Taste of China
127 Military St, (207) 532-1281
Chinese, Houlton
Eclipse Day Monday, 11am - 9pm

Featured Parks and Attractions

Aroostook State Park in Presque Isle, Maine, is the state's first state park. Visitors can enjoy hiking, bird watching, fishing, and canoeing on Echo Lake. The park's trail system includes a climb up Quaggy Jo Mountain, offering breathtaking views of the surrounding countryside. Campgrounds and picnic areas are also available. Aroostook State Park is a good spot for a day trip or an extended stay. Eclipse visitors will witness nearly three minutes of totality from this state park.

The Buffalo and Erie County Botanical Gardens are in a tri-domed Victorian-style glass conservatory inspired by the Crystal Palace in England. The gardens feature a rich collection of exotic plants from around the world arranged in both indoor and outdoor settings. Visitors can explore a tropical rainforest, a desert landscape, a traditional Japanese garden, and a children's garden. Celestial beauty takes center stage here at 3:18pm with nearly four minutes of totality during the eclipse.

Camel's Hump State Park, located in Vermont, is named after Camel's Hump, the third-highest peak in the state. Hikers will enjoy trails leading to the summit, which offer panoramic views of the Adirondacks. The park's forests, wetlands, and alpine tundra are home to a wide range of wildlife, making it an excellent location for bird watching and photography. Vermont's lush wilderness will be cast into shadow for two and a half minutes at this park on eclipse day.

Lake Francis State Park is in Pittsburgh, New Hampshire. Named after the scenic Lake Francis, the park is good choice for water activities such as fishing, boating, and swimming. Visitors can explore the park's forested surroundings on several hiking trails. Moose, deer, and a variety of birds and other wildlife populate the park. The park campgrounds and picnic areas are set along the shores of the lake with access to nearby trails. During the eclipse this 2,000 acre lake will be in the shadow of the moon for over three minutes.

Niagara Falls in New York is one of the most spectacular natural wonders in North America. It has three separate waterfalls – Horseshoe Falls, American Falls, and Bridal Veil Falls. Visitors can experience the falls up close on the famous Maid of the Mist boat tour or from the observation decks of Niagara Falls State Park, the oldest state park in the U.S. The park also features miles of hiking trails and historical exhibits. At 3:18pm the total solar eclipse commences, plunging the falls into darkness. What a great way to see Niagara Falls!

The Seneca Park Zoo in Rochester, New York, is home to many different species of animals, including African elephants, sea lions, snow leopards, and orangutans. One of the unique features of the zoo is the Animals of the Savanna exhibit, where guests can have an up-close encounter with giraffes. Be sure to check out the behavior of the savanna residents as the eclipse overtakes the area at 3:20pm. Wild!

The parks and attractions section in this book are organized by cities, which are highlighted in bold. Under each city, you will find a curated list of locations presented in alphabetical order. These locations may be within the city or situated within a 30-50 mile radius, encompassing the surrounding region. This method of organization allows for easy navigation and planning, ensuring you can make the most of your solar eclipse experience.

Never gaze at the sun without eye protection. Only remove eye protection during 100% totality.

Parks and Attractions

Jamestown, NY
Totality Starts at: 3:17pm EDT
Totality Lasts for: 2:52
Totality Ends at: 3:20pm EDT
NWS Office: www.weather.gov/buf
City Website: www.jamestownny.gov

Allegany State Park - Red House Beach
2373 ASP, US-1, Salamanca, NY 14779
parks.ny.gov/parks/73/
Totality Lasts for: 2:02

Allen Park
31 Hughes St
Jamestown, NY 14701
www.tourchautauqua.com

Chautauqua Gorge State Forest
Mayville, NY 14757
www.dec.ny.gov
Totality Lasts for: 3:33

Falconer Park
North Phetteplace Street
Falconer, NY 14733
falconerny.org

Fenton Historical Center
67 Washington St
Jamestown, NY 14701
www.fentonhistorycenter.org

Jackson-Taylor Park
10th &, Washington St
Jamestown, NY 14701
www.jamestownny.gov

Jamestown River Walk
203 W 2nd St
Jamestown, NY 14701
www.jamestownny.gov

Lakewood Beach
2 W Terrace Ave, Lakewood, NY 14750
Totality Starts at: 3:17pm EDT
Totality Lasts for: 3:02

Lake Erie State Park
5838 NY-5, Brocton, NY 14716
parks.ny.gov/parks/lakeerie
Totality Lasts for: 3:42

Long Point State Park on Lake Chautauqua
4459 NY-430
Bemus Point, NY 14712
parks.ny.gov/parks/109/

Lucille Ball Desi Arnaz Museum
2 W 3rd St
Jamestown, (716) 484-0800
lucy-desi.com

Lucille Ball Memorial Park & Statues
21 Boulevard Ave
Jamestown, NY 14701
www.tourchautauqua.com

Midway State Park
4859 NY-430, Bemus Point, NY 14712
parks.ny.gov/parks/167/
Totality Lasts for: 3:21

National Comedy Center
203 W 2nd St
Jamestown, NY 14701
comedycenter.org

Panama Rocks Scenic Park
11 Rock Hill Rd, Panama, NY 14767
www.panamarocks.com
Opens late spring

Paradise Bay Park Family Campground
2360 Shadyside Rd, Findley Lake, NY 14736
www.paradisebaypark.com
Totality Lasts for: 3:26

Dunkirk, NY
Totality Starts at: 3:17pm EDT
Totality Lasts for: 3:43
Totality Ends at: 3:21pm EDT
NWS Office: www.weather.gov/buf
City Website: www.cityofdunkirk.com

Bennett Beach
8276 Lake Shore Rd
Angola, NY 14006
www3.erie.gov/parks/bennett-beach

Canadaway Creek Nature Sanctuary
10836 Temple Rd
Dunkirk, NY 14048
www.cityofdunkirk.com

Cattaraugus Creek Harbor
Cattaraugus Creek
Irving, NY 14081
parks.ny.gov/parks/194/

Chautauqua Belle
78 Water St
Mayville, NY 14757
www.269belle.com

Dunkirk Boardwalk Market
8-22 Central Ave
Dunkirk, NY 14048
dunkirk-market.edan.io

Dunkirk City Pier
2 Central Ave
Dunkirk, NY 14048
www.cityofdunkirk.com

Dunkirk Lighthouse
1 Point Dr N
Dunkirk, NY 14048
www.dunkirklighthouse.com

Erie County Wendt Beach Park
8276 Old Lakeshore Rd
Angola, NY 14006
www3.erie.gov/parks/

Evangola State Park
10191 Old Lake Shore Rd, Irving, NY 14081
parks.ny.gov/parks/91/
Totality Lasts for: 3:45

Frank Lloyd Wright's Graycliff
6472 Old Lake Shore Rd
Derby, NY 14047
experiencegraycliff.org

Hamburg Beach/ Town Park
4420 Lake Shore Rd
Hamburg, NY 14075
www.hamburg-youth-rec-seniors.com

Lake Erie Beach Park
9568 Lake Shore Rd
Angola, NY 14006
www.townofevans.org

Lake Erie State Park
5838 NY-5, Brocton, NY 14716
parks.ny.gov/parks/lakeerie
Totality Lasts for: 3:42

Memorial Park
59 Lake Shore Dr W
Dunkirk, NY 14048
www.cityofdunkirk.com

Point Gratiot Park
NY-5 & Point Dr W
Dunkirk, NY 14048
www.tourchautauqua.com

Solé at Woodlawn Beach
3580 Lake Shore Rd
Blasdell, NY 14219
www.soleatwoodlawnbeach.com

Sturgeon Point Marina
618 Sturgeon Point Rd
Derby, NY 14047
www.townofevans.org

Sunset Bay State Marine Park
12952 Allegany Rd
Irving, NY 14081
parks.ny.gov/parks/194/

Westfield / Lake Erie KOA Journey
8001 East Lake Road, NY-5
Westfield, NY 14787
koa.com/campgrounds/westfield/

Never gaze at the sun without eye protection. Only remove eye protection during 100% totality.

Buffalo, NY
Totality Starts at: 3:18pm EDT
Totality Lasts for: 3:45
Totality Ends at: 3:22pm EDT
NWS Office: www.weather.gov/buf
City Website: www.buffalony.gov

Amherst State Park
390 Mill St, Buffalo, NY 14221
www.amherststatepark.org
Totality Lasts for: 3:43

Broderick Park
1170 Niagara St
Buffalo, NY 14213
www.buffalony.gov

Buffalo AKG Art Museum
1285 Elmwood Ave
Buffalo, NY 14222
buffaloakg.org

Buffalo and Erie County Botanical Gardens
2655 South Park Ave
Buffalo, NY 14218
www.buffalogardens.com

Buffalo Harbor State Park
1111 Fuhrmann Boulevard, Buffalo, NY 14203
parks.ny.gov/parks/191/
Totality Lasts for: 3:45

Buffalo Museum of Science
1020 Humboldt Pkwy
Buffalo, NY 14211
www.sciencebuff.org

Buffalo Naval Park
1 Naval, Marina Park S
Buffalo, NY 14202
buffalonavalpark.org

Buffalo Transportation Pierce Arrow Museum
263 Michigan Ave
Buffalo, NY 14203
www.pierce-arrow.com

Canalside
44 Prime St
Buffalo, NY 14202
www.buffalowaterfront.com

Cazenovia Park
Warren Spahn Way
Buffalo, NY 14220
www.bfloparks.org/parks/cazenovia-park/

Charles E. Burchfield Nature & Art Center
2001 Union Rd
West Seneca, NY 14224
www.burchfieldnac.org

Como Lake Park
2220 Como Park Blvd
Lancaster, NY 14086
www3.erie.gov/parks/

Darien Lakes State Park
10475 Harlow Rd, Darien Center, NY 14040
parks.ny.gov/parks/144/
Totality Lasts for: 3:44

Delaware Park
84 Parkside Ave
Buffalo, NY 14214
www.bfloparks.org

Frank Lloyd Wright's Martin House
125 Jewett Pkwy
Buffalo, NY 14214
martinhouse.org

Glen Park
5565 Main St
Williamsville, NY 14221
walkablewilliamsville.com

Herschell Carrousel Factory Museum
180 Thompson St
North Tonawanda, NY 14120
www.carrouselmuseum.org

Hoyt Lake
199 Lincoln Pkwy
Buffalo, NY 14222
www.bfloparks.org

Japanese Garden
1 Museum Ct
Buffalo, NY 14216
www.bfloparks.org

Knox Farm State Park
437 Buffalo Rd, East Aurora, NY 14052
parks.ny.gov/parks/163/
Totality Lasts for: 3:43

Lafayette Square
415 Main St
Buffalo, NY 14203
www.buffalony.gov

Peace Bridge
1 Peace Bridge
Buffalo, NY 14213
www.peacebridge.com

Stiglmeier Park
810 Losson Rd
Cheektowaga, NY 14227
www.tocny.org

The Buffalo Zoo
300 Parkside Ave
Buffalo, NY 14214
www.buffalozoo.org

The Ralph C. Wilson, Jr. Children's Museum
130 Main St
Buffalo, NY 14202
www.exploreandmore.org

Theodore Roosevelt Inaugural Historic Site
641 Delaware Ave
Buffalo, NY 14202
www.trsite.org

Wilkeson Pointe
225 Fuhrmann Boulevard
Buffalo, NY 14203
www.buffalowaterfront.com

Niagara Falls, NY
Totality Starts at: 3:18pm EDT
Totality Lasts for: 3:28
Totality Ends at: 3:21pm EDT
NWS Office: www.weather.gov/buf
niagarafallsusa.com

Aquarium of Niagara
701 Whirlpool St
Niagara Falls, NY 14301
www.aquariumofniagara.org

Buckhorn Island State Park
E River Rd, Grand Island, NY 14072
parks.ny.gov/parks/buckhornisland/
Totality Lasts for: 3:33

Cave of the Winds
Goat Island Rd
Niagara Falls, NY 14303
www.niagarafallsstatepark.com

Devil's Hole State Park
Niagara Scenic Pkwy
Niagara Falls, NY 14305
parks.ny.gov/parks/42/

Fort Niagara State Park
1 Scott Ave
Youngstown, NY 14174
parks.ny.gov/parks/175/

Four Mile Creek State Park
1055 Lake Rd
Youngstown, NY 14174
parks.ny.gov/parks/6/

Golden Hill State Park
9691 Lower Lake Rd
Barker, NY 14012
parks.ny.gov/parks/goldenhill/

Hyde Park
3200 Pine Ave
Niagara Falls, NY 14301
niagarafallsusa.org

Maid of the Mist
1 Prospect St
Niagara Falls, NY 14303
www.maidofthemist.com

Niagara Falls State Park
332 Prospect St
Niagara Falls, NY 14303
www.niagarafallsstatepark.com

Old Fort Niagara
102 Morrow Plaza
Youngstown, NY 14174
www.oldfortniagara.org

Reservoir State Park
5777 Witmer Rd
Niagara Falls, NY 14305
parks.ny.gov/parks/75/

Whirlpool State Park
Niagara Scenic Pkwy
Niagara Falls, NY 14303
parks.ny.gov/parks/105/

Never gaze at the sun without eye protection. Only remove eye protection during 100% totality.

Wilson-Tuscarora State Park
3371 W Lake Rd
Wilson, NY 14172
parks.ny.gov/parks/wilsontuscarora/

Batavia, NY
Totality Starts at: 3:19pm EDT
Totality Lasts for: 3:43
Totality Ends at: 3:23pm EDT
NWS Office: www.weather.gov/buf
www.batavianewyork.com

Akron Falls Park
44 Parkview Dr
Akron, NY 14001
www.erie.gov/parks/

Centennial Park
151 State St
Batavia, NY 14020
www.batavianewyork.com

Darien Lakes State Park
10475 Harlow Rd
Darien Center, NY 14040
parks.ny.gov/parks/144/

DeWitt Recreation Area
115 Cedar St
Batavia, NY 14020
www.batavianewyork.com

Genesee County Park & Forest
11095 Bethany Center Rd, East Bethany
Totality Starts at: 3:19pm EDT
Totality Lasts for: 3:35

Letchworth State Park
1 Letchworth State Park, Castile, NY 14427
parks.ny.gov/parks/79/
Totality Lasts for: 3:00

Silver Lake State Park
4229 W Lake Rd, Castile, NY 14550
parks.ny.gov/parks/silverlake/
Totality Lasts for: 3:08

Stony Brook State Park
10820 NY-36, Dansville, NY 14437
parks.ny.gov/parks/stonybrook/
Totality Lasts for: 1:54

Rochester, NY
Totality Starts at: 3:20pm EDT
Totality Lasts for: 3:40
Totality Ends at: 3:23pm EDT
NWS Office: www.weather.gov/buf
City Website: www.visitrochester.com

Abraham Lincoln Park
Empire Blvd
Webster, NY 14580
www.monroecounty.gov/parks-bayeast

Beechwood State Park
Lake Rd
Sodus, NY 14551
townofsodus.net/camp-beechwood

Braddock Bay Park
199 E Manitou Rd
Rochester, NY 14612
greeceny.gov/braddock-bay-park

Cayuga Lake State Park
2678 Lower Lake Rd, Seneca Falls, NY 13148
parks.ny.gov/parks/cayugalake/
Totality Lasts for: 2:07

Durand Eastman Beach
1342 Lakeshore Blvd
Rochester, NY 14617
www.cityofrochester.gov/durandbeach/

George Eastman Museum
900 East Ave
Rochester, NY 14607
eastman.org

Hamlin Beach State Park
1 Hamlin Beach State Park, Hamlin, NY 14464
parks.ny.gov/parks/20/
Totality Lasts for: 3:37

Highland Park
180 Reservoir Ave
Rochester, NY 14620
www.monroecounty.gov/parks-highland

Irondequoit Bay State Marine Park
Culver Rd, Irondequoit, NY 14622
Totality Starts at: 3:20pm EDT
Totality Lasts for: 3:42

Lakeside State Park
NY-18, Waterport, NY 14571
parks.ny.gov/parks/161/
Totality Lasts for: 3:30

Memorial Art Gallery
500 University Ave
Rochester, NY 14607
mag.rochester.edu

Ontario Beach Park
50 Beach Ave
Rochester, NY 14612
www.cityofrochester.gov/ontariobeachpark/

RMSC Rochester Museum & Science Center
657 East Ave
Rochester, NY 14607
www.rmsc.org

Seneca Lake State Park
1 Lake Front Dr, Geneva, NY 14456
parks.ny.gov/parks/125/
Totality Lasts for: 2:19

Seneca Park Zoo
2222 St Paul St
Rochester, NY 14621
senecaparkzoo.org

Sodus Bay Historical Society
7606 N Ontario St
Sodus Point, NY 14555
www.sodusbaylighthouse.org

Sodus Point Beach Park
7958 Wickham Blvd
Sodus Point, NY 14555
soduspoint.info/beach/

Susan B. Anthony Museum & House
17 Madison St
Rochester, NY 14608
www.susanb.org

The Strong National Museum of Play
1 Manhattan Square Dr
Rochester, NY 14607
www.museumofplay.org

Turning Point Park
260 Boxart St
Rochester, NY 14612
www.cityofrochester.gov/turningpoint/

Webster County Park
Webster, NY 14580
Totality Starts at: 3:20pm EDT
Totality Lasts for: 3:41

Oswego, NY
Totality Starts at: 3:21pm EDT
Totality Lasts for: 3:30
Totality Ends at: 3:25pm EDT
NWS Office: www.weather.gov/buf
City Website: townofoswego.com

Breitbeck Park
91 Lake St
Oswego, NY 13126
www.oswegony.org

Children's Museum of Oswego (CMOO)
7 W Bridge St
Oswego, NY 13126
www.cmoo.org

Chimney Bluffs State Park
7700 Garner Rd, Wolcott, NY 14590
parks.ny.gov/parks/43/
Totality Lasts for: 3:28

Fair Haven Beach State Park
14985 State Park Rd, Fair Haven, NY 13064
parks.ny.gov/parks/12/
Totality Lasts for: 3:25

H. Lee White Maritime Museum
1 W 1st St
Oswego, NY 13126
www.hlwmm.org

Lake Bluff Campground
7150 Garner Rd
Wolcott, NY 14590
www.lakebluffrvpark.com

Mexico Point State Park
County Rte 40, Mexico, NY 13114
mexicopointpark.com
Totality Lasts for: 3:27

Salmon River Falls Unique Area
Falls Rd, Richland, NY 13144
www.dec.ny.gov/lands/63578.html
Totality Lasts for: 3:17

Never gaze at the sun without eye protection. Only remove eye protection during 100% totality.

Sandy Island Beach State Park
3387 County Rte 15, Pulaski, NY 13142
parks.ny.gov/parks/153/
Totality Lasts for: 3:34

Selkirk Shores State Park
7101 State Rte 3, Pulaski, NY 13142
parks.ny.gov/parks/84/
Totality Lasts for: 3:29

Sterling Creek Campground
14983 Juniper Hill Rd
Sterling, NY 13156
sterlingcreekcampground.com

Sterling Nature Center
15730 Jensvold Rd
Sterling, NY 13156
www.cayugacounty.us/446/

Sunset Bay Park
144 Lake Rd
Oswego, NY 13126
www.oswegony.org

Syracuse, NY
Totality Starts at: 3:22pm EDT
Totality Lasts for: 1:45
Totality Ends at: 3:24pm EDT
NWS Office: www.weather.gov/bgm
City Website: www.visitsyracuse.com

Beaver Lake Nature Center
8477 E Mud Lake Rd, Baldwinsville, NY 13027
www.onondagacountyparks.com
Totality Lasts for: 2:46

Erie Canal Museum
318 Erie Blvd E
Syracuse, NY 13202
eriecanalmuseum.org

Everson Museum of Art
401 Harrison St
Syracuse, NY 13202
www.everson.org

Long Branch Park
3813 Long Branch Rd, Liverpool, NY 13090
Totality Starts at: 3:22pm EDT
Totality Lasts for: 2:15

Lysander Town Park
Baldwinsville, NY 13027
www.townoflysander.org
Totality Lasts for: 2:40

Museum of Science & Technology
500 S Franklin St
Syracuse, NY 13202
www.most.org

Onondaga Lake Park
106 Lake Dr, Liverpool, NY 13088
www.onondagacountyparks.com
Totality Lasts for: 2:03

Battle Island State Park Public Golf Course
2150 NY-48, Fulton, NY 13069
parks.ny.gov/golf/5/
Totality Lasts for: 3:17

Van Buren Central Park
7350 Canton St, Baldwinsville, NY 13027
townofvanburen.com
Totality Lasts for: 2:26

William J. Farley Jr. Community Park
132-198 Chestnut St, Phoenix, NY 13135
Totality Starts at: 3:22pm EDT
Totality Lasts for: 2:46

Watertown, NY
Totality Starts at: 3:22pm EDT
Totality Lasts for: 3:38
Totality Ends at: 3:26pm EDT
NWS Office: www.weather.gov/buf
City Website: www.watertown-ny.gov

Boldt Castle & Boldt Yacht House
1 Heart Island
Alexandria Bay, (315) 482-9724
www.boldtcastle.com

Burnham Point State Park
340765 NY-12E, Cape Vincent, NY 13618
parks.ny.gov/parks/burnhampoint/
Totality Lasts for: 3:19

Cedar Point State Park
36661 State Park Dr, Clayton, NY 13624
parks.ny.gov/parks/cedarpoint/
Totality Lasts for: 3:17

Higley Flow State Park
442 Cold Brook Dr, Colton, NY 13625
parks.ny.gov/parks/58/
Totality Lasts for: 3:27

Historic Thompson Park
Thompson Park
Watertown, NY 13601
historicthompsonpark.org

Jacques Cartier State Park
NY-12, Morristown, NY
www.stateparks.com/jacques_cartier.html
Totality Lasts for: 2:52

Keewaydin State Park
45165 NY-12, Alexandria Bay, NY 13607
parks.ny.gov/parks/keewaydin/
Totality Lasts for: 3:14

Kring Point State Park
25950 Kring Point Rd, Redwood, NY 13679
parks.ny.gov/parks/14/
Totality Lasts for: 3:10

Natural Bridge / Watertown KOA Journey
6081 State Rte 3
Natural Bridge, NY 13665
koa.com/campgrounds/watertown/

Southwick Beach State Park
8119 Southwicks Pl, Henderson, NY 13650
parks.ny.gov/parks/36/
Totality Lasts for: 3:39

Robert G. Wehle State Park
5182 State Park Rd, Henderson, NY 13650
parks.ny.gov/parks/robertwhele/
Totality Lasts for: 3:38

Sackets Harbor Battlefield
504 W Main St
Sackets Harbor, NY 13685
parks.ny.gov/historic-sites/

Watertown Public Square
1 Public Square
Watertown, NY 13601
www.watertown-ny.gov

Wellesley Island State Park
44927 Cross Island Rd, Fineview, NY 13640
parks.ny.gov/parks/52
Totality Lasts for: 3:10

Westcott Beach State Park
State Rte 3, Henderson, NY 13650
parks.ny.gov/parks/westcottbeach/
Totality Lasts for: 3:38

Whetstone Gulf State Park
6065 W Rd, Lowville, NY 13367
parks.ny.gov/parks/whetstonegulf/
Totality Lasts for: 3:14

Zoo New York
1 Thompson Park
Watertown, NY 13601
www.zoonewyork.org

Plattsburgh, NY
Totality Starts at: 3:25pm EDT
Totality Lasts for: 3:34
Totality Ends at: 3:29pm EDT
NWS Office: www.weather.gov/bvt
www.cityofplattsburgh-ny.gov

Ausable Chasm
2144 US-9, Ausable Chasm, NY 12911
ausablechasm.com
Totality Lasts for: 3:27

Ausable Chasm Campground
634 NY-373
Keeseville, NY 12944
ausablechasm.com

Cobble Lookout Trailhead
Co Hwy 18A
Wilmington, NY 12997
pureadirondacks.com

Lake Placid Olympic Museum
2634 Main St, Lake Placid, (518) 523-1655
lakeplacidolympicmuseum.org
Totality Lasts for: 3:22

MacDonough Park
42 City Hall Pl
Plattsburgh, NY 12901
www.plattsburghrecreation.com

Macomb Reservation State Park
201 Campsite Rd, Schuyler Falls, NY 12985
parks.ny.gov/parks/macombreservation/
Totality Lasts for: 3:34

Never gaze at the sun without eye protection. Only remove eye protection during 100% totality.

Point Au Roche State Park
19 Camp Red Cloud Rd, Plattsburgh, NY 12901
Totality Starts at: 3:25pm EDT
Totality Lasts for: 3:33

Samuel Champlain Monument Park
30 Cumberland Ave
Plattsburgh, NY 12901
www.plattsburghrecreation.com

Vermont

St Albans City, VT
Totality Starts at: 3:26pm EDT
Totality Lasts for: 3:33
Totality Ends at: 3:29pm EDT
NWS Office: www.weather.gov/bvt
City Website: www.stalbansvt.com

Alburgh Dunes State Park
151 Coon Point Rd, Alburgh, VT 05440
vtstateparks.com/alburgh.html
Totality Lasts for: 3:31

Grand Isle State Park
36 E Shore S, Grand Isle, VT 05458
vtstateparks.com/grandisle.html
Totality Lasts for: 3:33

Hard'ack Recreation Area
264 Hard'ack
St Albans City, VT 05478
stalbansvt.myrec.com

Kill Kare State Park
2714 Hathaway Point Rd, St. Albans Town, VT
vtstateparks.com/killkare.html
Totality Lasts for: 3:33

Knight Point State Park
44 Knights Point Rd, North Hero, VT 05474
vtstateparks.com/knightpoint.html
Totality Lasts for: 3:33

Lake Carmi State Park
460 Marsh Farm Rd, Franklin, VT 05457
vtstateparks.com/carmi.html
Totality Lasts for: 3:31

Missisquoi Valley Rail Trail
St Albans City, VT 05478
www.mvrailtrail.org
Totality Lasts for: 3:33

North Hero State Park
3803 Lakeview Dr, North Hero, VT 05474
vtstateparks.com/northhero.html
Totality Lasts for: 3:30

St Albans Bay Town Park
596 Lake St, St.
Albans Town, VT 05478
www.stalbanstown.com

Burlington, VT
Totality Starts at: 3:26pm EDT
Totality Lasts for: 3:13
Totality Ends at: 3:29pm EDT
NWS Office: www.weather.gov/bvt
City Website: www.burlingtonvt.gov

Battery Park
Battery Park Extension
Burlington, VT 05401
enjoyburlington.com/venue/battery-park/

Birds of Vermont Museum
900 Sherman Hollow Rd, Huntington, VT 05462
www.birdsofvermont.org
Totality Lasts for: 2:47

Button Bay State Park
5 Button Bay State Park Rd, Vergennes
Totality Starts at: 3:26pm EDT
Totality Lasts for: 2:34

Causeway Park
781 Blakely Rd
Colchester, VT 05446
Totality Lasts for: 3:25

Church Street Marketplace
131 Church St, Burlington, VT 05401
Shopping, Burlington
Eclipse Day Monday, 10 AM - 7 PM

D.A.R. State Park
6750 VT-17, Addison, VT 05491
vtstateparks.com/dar.html
Totality Lasts for: 2:05

ECHO, Leahy Center for Lake Champlain
1 College St
Burlington, VT 05401
www.echovermont.org

Ethan Allen Homestead Museum and Historic Site
1 Ethan Allen Homestead
Burlington, VT 05408
www.ethanallenhomestead.org

Ethan Allen Park
1006 North Ave
Burlington, VT 05408
enjoyburlington.com/venue/ethan-allen-park/

Kingsland Bay State Park
787 KB State Park Rd, Ferrisburgh
vtstateparks.com/kingsland.html
Totality Lasts for: 2:43

Leddy Park
216 Leddy Park Rd
Burlington, VT 05408
enjoyburlington.com/place/leddy-park-3/

Maple Street Park and Pool
75 Maple St
Essex Junction, VT 05452
www.ejrp.org

Mount Mansfield State Forest
175 Pleasant Valley Rd, Underhill
fpr.vermont.gov/mt-mansfield-state-forest-0
Totality Lasts for: 2:58

Mt. Philo State Park
5425 Mt Philo Rd, Charlotte
vtstateparks.com/philo.html
Totality Lasts for: 2:47

North Beach Park
Burlington, VT 05401
enjoyburlington.com/venue/north-beach/
Totality Lasts for: 3:17

Niquette Bay State Park
274 Raymond Rd, Colchester, VT 05446
vtstateparks.com/niquette.html
Totality Lasts for: 3:25

Oakledge Park
11 Flynn Ave
Burlington, VT 05401
enjoyburlington.com/place/oakledge-park-2/

Overlook Park
1575 Spear St
South Burlington, VT 05403
www.southburlingtonvt.gov

Red Rocks Park
4 Central Ave
South Burlington, VT 05403
www.southburlingtonvt.gov

Sand Bar State Park
1215 U.S. Rte 2, Milton, VT 05468
vtstateparks.com/sandbar.html
Totality Lasts for: 3:29

Shelburne Museum Covered Bridge
5555 Shelburne Rd
Shelburne, VT 05482
shelburnemuseum.org

Smugglers' Notch State Park Campground
6443 Mountain Rd, Stowe, VT 05672
www.vtstateparks.com
Totality Lasts for: 3:01

Underhill State Park
352 Mountain Rd, Underhill, VT 05489
vtstateparks.com/underhill.html
Totality Lasts for: 3:06

Vermont Teddy Bear Factory
6655 Shelburne Rd, Shelburne, (802) 985-3001
www.vermontteddybear.com
Totality Lasts for: 3:00

Waterfront Park
20 Lake St
Burlington, VT 05401
enjoyburlington.com/place/waterfront-park/

Montpelier, VT
Totality Starts at: 3:27pm EDT
Totality Lasts for: 1:42
Totality Ends at: 3:29pm EDT
NWS Office: www.weather.gov/bvt
City Website: www.montpelier-vt.org

Camel's Hump State Park
3429 Camels Hump Rd, Duxbury, VT 05676
www.vtstateparks.com/camelshump.html
Totality Lasts for: 2:31

Never gaze at the sun without eye protection. Only remove eye protection during 100% totality.

Cold Hollow Cider Mill
3600 Waterbury-Stowe Rd, Waterbury Center
www.coldhollow.com
Totality Lasts for: 2:34

Hubbard Park
400 Parkway St
Montpelier, VT 05602
www.montpelier-vt.org/238/Hubbard-Park

Kenneth Ward Park
4806 VT-100B, Middlesex, VT 05602
Totality Starts at: 3:27pm EDT
Totality Lasts for: 2:09

Little River State Park
3444 Little River Rd, Waterbury Village HD
vtstateparks.com/littleriver.html
Totality Lasts for: 2:40

Morse Farm Maple Sugarworks
1168 County Rd, Montpelier, VT 05602
www.morsefarm.com
Totality Lasts for: 1:47

North Branch Nature Center
713 Elm St
Montpelier, VT 05602
www.northbranchnaturecenter.org

Vermont Historical Society Museum
109 State St
Montpelier, VT 05609
www.vermonthistory.org

Vermont State House
115 State St
Montpelier, VT 05633
statehouse.vermont.gov

Waterbury Center State Park
177 Reservoir Rd, Waterbury Center
vtstateparks.com/waterbury.html
Totality Lasts for: 2:34

New Hampshire

Colebrook, NH
Totality Starts at: 3:28pm EDT
Totality Lasts for: 2:59
Totality Ends at: 3:31pm EDT
NWS Office: www.weather.gov/gyx
City Website: colebrooknh.org

Beaver Brook Falls Wayside State Park
432 NH-145, Colebrook, NH 03576
Totality Starts at: 3:28pm EDT
Totality Lasts for: 3:01

Coleman State Park
1166 Diamond Pond Rd, Stewartstown
Totality Starts at: 3:28pm EDT
Totality Lasts for: 2:59

Columbia Covered Bridge
Columbia Bridge Rd, North Stratford
www.nh.gov/nhdhr/bridges/p63.html
Totality Lasts for: 2:55

Connecticut Lakes State Forest
Pittsburg, NH 03592
www.nhstateparks.org
Totality Lasts for: 3:26

Lake Francis State Park
439 River Rd, Pittsburg, NH 03592
www.nhstateparks.org
Totality Lasts for: 3:12

Dixville Notch State Park
NH-26, Colebrook, NH 03576
Totality Starts at: 3:28pm EDT
Totality Lasts for: 2:45

Northumberland, NH
Totality Starts at: 3:28pm EDT
Totality Lasts for: 1:49
Totality Ends at: 3:30pm EDT
NWS Office: www.weather.gov/gyx
www.northumberlandnh.org

Androscoggin Wayside Park
Route 16, Milan, NH 03588
Totality Starts at: 3:29pm EDT
Totality Lasts for: 2:07

Crystal Falls
96 Dewey Hill Rd
Stark, NH 03582
Totality Lasts for: 1:33

Mollidgewock State Park
1437 Berlin Rd, Errol, NH 03579
Totality Starts at: 3:29pm EDT
Totality Lasts for: 2:06

Stark Covered Bridge
NH-110, Groveton, NH 03582
www.nh.gov/nhdhr/bridges/p71.html
Totality Lasts for: 1:35

Umbagog Lake State Park Campground
RR 26, Errol, NH 03579
Totality Starts at: 3:29pm EDT
Totality Lasts for: 1:34

Maine

Houlton, ME
Totality Starts at: 3:32pm EDT
Totality Lasts for: 3:20
Totality Ends at: 3:35pm EDT
NWS Office: www.weather.gov/car
www.houlton-maine.com

Aroostook State Park
87 State Park Rd, Presque Isle, ME 04769
Totality Starts at: 3:32pm EDT
Totality Lasts for: 2:56

Baxter State Park
Millinocket, ME 04462
Totality Starts at: 3:30pm EDT
Totality Lasts for: 3:23

Birch Point Lodge Campgrounds
33 Birch Point Lane, Island Falls, ME 04747
birchpointcampground.com
Totality Lasts for: 3:20

Houlton Riverfront Park
49 North St
Houlton, ME 04730
www.houlton-maine.com

Lily Bay State Park
425 Lily Bay Rd, Beaver Cove, ME 04441
Totality Starts at: 3:30pm EDT
Totality Lasts for: 3:13

Mattawamkeag Wilderness Park
1513 Wilderness Park Rd, Mattawamkeag
Totality Starts at: 3:32pm EDT
Totality Lasts for: 2:04

Mt Katahdin
Northeast Piscataquis, ME 04462
Totality Starts at: 3:30pm EDT
Totality Lasts for: 3:24

Peaks-Kenny State Park
401 State Park Rd, Dover-Foxcroft, ME 04426
Totality Starts at: 3:31pm EDT
Totality Lasts for: 2:05

South Branch Pond Campground
Baxter State Park, Millinocket, ME 04462
Totality Starts at: 3:30pm EDT
Totality Lasts for: 3:21

Never gaze at the sun without eye protection. Only remove eye protection during 100% totality.

Resources

National Weather Service Offices Covering The Path Of Totality For The April 8, 2024 Solar Eclipse

During the 2017 eclipse, many National Weather Service offices established dedicated a web page about the eclipse and the weather conditions surrounding the event. Keep these sites in mind and check with them as the event gets closer.

For help with using the National Weather Service website, visit this website: www.weather.gov/contact

NWS Austin/San Antonio, TX
Office ID: EWX
2090 Airport Road
New Braunfels, TX 78130
Phone: (830) 629-0130
Web: *www.weather.gov/ewx*

NWS San Angelo, TX
Office ID: SJT
7654 Knickerbocker Road
San Angelo, TX 76904
Phone: (325) 944-9445
Web: *www.weather.gov/sjt*

NWS Fort Worth/Dallas, TX
Office ID: FWD
3401 Northern Cross Boulevard
Fort Worth, Texas 76137
Phone: (817) 429-2631
Web: *www.weather.gov/fwd*

NWS Houston/Galveston, TX
Office ID: HGX
1353 FM 646 Suite 202
Dickinson, TX 77539
Phone: (281) 337-5074
Web: *www.weather.gov/hgx*

NWS Shreveport, LA
Office ID: SHV
5655 Hollywood Ave.
Shreveport, LA 71109
Phone: 318-631-3669
Web: *www.weather.gov/shv*

NWS Little Rock, AR
Office ID: LZK
8400 Remount Road
North Little Rock, AR 72118
Phone: (501) 834-0308
Web: *www.weather.gov/lzk*

NWS Memphis, TN
Office ID: MEG
7777 Walnut Grove Road, OM1
Memphis, TN 38120
Phone: (901) 544-0399
Web: *www.weather.gov/meg*

NWS Springfield, MO
Office ID: SGF
5805 West Highway EE
Springfield, MO 65802-8430
Phone: Recording: (417) 869-4491
Web: *www.weather.gov/sgf*

NWS St. Louis, MO
Office ID: LSX
12 Missouri Research Park Drive
St. Charles, MO 63304-5685
Phone: (636) 441-8467
Web: *www.weather.gov/lsx*

NWS Paducah, KY
Office ID: PAH
8250 Kentucky Highway 3520
West Paducah, KY 42086-9762
Phone: (270) 744-6440
Web: *www.weather.gov/pah*

NWS Central Illinois
Office ID: ILX
1362 State Route 10
Lincoln, IL 62656
Phone: (217) 732-7321
Web: *www.weather.gov/ilx*

NWS Indianapolis, IN
Office ID: IND
6900 West Hanna Avenue
Indianapolis, IN 46241-9526
Phone: (317) 856-0664
Web: *www.weather.gov/ind*

NWS Louisville, KY
Office ID: LMK
6201 Theiler Lane
Louisville, KY 40229-1476
Phone: 502-969-8842
Web: *www.weather.gov/lmk*

NWS Northern Indiana
Office ID: IWX
7506 E 850 N
Syracuse, IN 46567
Phone: (574) 834-1104
Web: *www.weather.gov/iwx*

NWS Wilmington, OH
Office ID: ILN
1901 South State Route 134
Wilmington, OH 45177
Phone: (937) 383-0031
Web: *www.weather.gov/iln*

NWS Cleveland, OH
Office ID: CLE
925 Keynote Circle
Brooklyn Heights, OH 44131
Phone: (216) 416-2900
Web: *www.weather.gov/cle*

NWS Pittsburgh, PA
Office ID: PBZ
192 Shafer Road
Moon Township, PA 15108
Phone: Public (412) 262-2170
Web: *www.weather.gov/pbz*

NWS Buffalo, NY
Office ID: BUF
587 Aero Drive
Cheektowaga, NY 14225
Phone: (716) 565-0204
Web: *www.weather.gov/buf*

NWS State College, PA
Office E ID: CTP
328 Innovation Blvd, Suite 330
State College, PA 16803
Phone: (814) 954-6440
Web: *www.weather.gov/ctp*

NWS Binghamton, NY
Office ID: BGM
32 Dawes Drive
Johnson City, NY 13790
Phone: (607) 729-1597
Web: *www.weather.gov/bgm*

NWS Burlington, VT
Office ID: BTV
1200 Airport Drive
South Burlington, VT 05403
Phone: (802) 862-2475
Web: *www.weather.gov/bvt*

NWS Albany, NY
Office ID: ALY
1400 Washington Avenue
Albany, NY 12222
Phone: (518) 626-7570
Web: *www.weather.gov/aly*

NWS Gray - Portland, ME
Office ID: GYX
P.O. Box 1208
1 Weather Lane
Gray, ME 04039
Phone: (207) 688-3216
Web: *www.weather.gov/gyx*

NWS Caribou, ME
Office ID: CAR
810 Main Street
Caribou, ME 04736
Phone: (207) 492-0182 (person)
Web: *www.weather.gov/car*

Weather Related Websites

www.cleardarksky.com
ClearDarkSky is a website that offers a variety of tools including Clear Sky Charts, which provide highly accurate forecasts of astronomical observing conditions and cloud cover for over 6300 locations in North America. Click on Clear Sky Charts in the upper left to get started.

clearoutside.com/forecast/31.43/-97.72 (opens at Gatesville, Texas)
Clearoutside is a website that provides a graphical interface to a 7 day clear sky forecast. After selecting a location you are shown an hour by hour graph indicating the % of sky obscured by total clouds, low clouds, medium clouds, high clouds and more.

weather.cod.edu/#
The College of DuPage (COD) website provides access to weather analysis tools such as satellite and radar, numerical models, text products, and data to assist in understanding current weather forecasts. The College of DuPage (COD) Meteorology program, known as NexLab, provides students with a comprehensive meteorology education, offering a wide range of courses in forecasting, severe weather, and traditional atmospheric sciences.

eclipsophile.com
Eclipsophile.com is a website by Jay Anderson dedicated to providing climate and weather information for celestial events.

worldview.earthdata.nasa.gov
NASA's Earth Observing System Data and Information System (EOSDIS) allows users to interactively browse and download data from over 1000 global, full-resolution satellite imagery layers. Many of these layers are updated daily and accessible within three hours of observation, offering near real-time views of Earth. Geostationary imagery layers are available in ten-minute increments for the past 90 days, enabling near real-time monitoring of global changes. EOSDIS supports browsing on mobile devices, allowing easy access to imagery on tablets and smartphones.

www.star.nesdis.noaa.gov/GOES/
The GOES Imagery Viewer website, provided by NOAA/NESDIS/STAR, offers users access to images and animations from the GOES (Geostationary Operational Environmental Satellite) series of satellites.

digital.weather.gov
Sky Cover Forecast
The National Weather Service's Graphical Forecast website (digital.weather.gov) provides graphical forecasts based on data from the National Digital Forecast Database. From the main page, select Sky Cover in the upper left to display the current sky cover forecast.

spotwx.com
SpotWx is a weather forecasting tool that enables users to select any location on the globe and receive detailed weather predictions. The website provides multiple weather models (NAM, GFS, HRRR...) with varying forecast lengths and accuracy, allowing users to choose the model that best covers their area of interest and study the charts.

zoom.earth
Zoom Earth is an interactive world weather map using satellite images from the GOES series of satellites. Shows live and past images with a convenient interface for zooming in and scrolling previous images.

Never gaze at the sun without eye protection. Only remove eye protection during 100% totality.

Eclipse Related Websites

American Astronomical Society - *eclipse.aas.org/eclipse-america-2024*

Eclipse Over Texas 2024 - *eclipseovertexas2024.com*

Eclipse2024.org (by Dan McGlaun) - *eclipse2024.org*

EclipseWise.com (by Fred Espenak) - *eclipsewise.com*

Eclipsophile.com (by Jay Anderson) - *eclipsophile.com*

Enjoy Illinois: Makanda Solar Eclipse - *www.enjoyillinois.com/explore/listing/solar-eclipse-2024/*

Experience the Eclipse in Arkansas - *www.arkansas.com/things-to-do/outdoors/skygazing/2024-eclipse*

Eye Safety During Solar Eclipses - *eclipse.gsfc.nasa.gov/SEhelp/safety2.html*

GreatAmericanEclipse.com (by Michael Zeiler) - *www.greatamericaneclipse.com/april-8-2024*

Hermit Eclipse (by Ian Cameron Smith) - *moonblink.info/Eclipse/eclipse/2024_04_08*

Hill Country Alliance Eclipse Portal - *hillcountryalliance.org/eclipse/*

Hot Springs: The Place to Be for Eclipse 2024 - *www.totaleclipsearkansas.com*

Indiana Total Solar Eclipse - *www.in.gov/dnr/places-to-go/events/2024-solar-eclipse/*

Kerrville Eclipse - *www.kerrvilletexascvb.com/kerrvilletxeclipse*

Maine Eclipse - *maineeclipse.com*

Missouri Eclipse - *moeclipse.org*

NASA Science: Eclipses - *solarsystem.nasa.gov/eclipses/2024/apr-8-total/overview/*

New Hampshire Total Solar Eclipse - *www.visitnh.gov/industry-members/work-together/total-solar-eclipse*

Ohio Total Solar Eclipse - *eclipse.ohio.gov*

Ozark Gateway Region Total Eclipse - *ozarkgateway.com/2024-arkansas-solar-eclipse/*

Poplar Bluff Total Solar Eclipse - *visitbutlercountymo.com/2024-eclipse/*

Protect Your Eyes from the Sun! - *preventblindness.org/solar-eclipse-and-your-eyes/*

Rochester Total Solar Eclipse - *rochestereclipse2024.org*

Solar Eclipses and Maps (by Xavier M. Jubier) - *xjubier.free.fr*

Solar Eclipse Timer App (by Gordon Teleppun) - *www.solareclipsetimer.com*

Southern Illinois University: Solar Eclipse Crossroads - *eclipse.siu.edu*

West Plains Total Solar Eclipse - *explorewestplains.com/tse/*

Glossary of Eclipse Terms

Annularity: A term used to describe the appearance of a solar eclipse when the Moon's apparent size is smaller than the Sun's, creating a ring-like or annular shape of the Sun's outer edge during the event. The next annular solar eclipse will occur on October 14, 2023, and will be visible from parts of North America.

Baily's Beads: The phenomenon observed just before and after totality during a solar eclipse, where small, bright spots of sunlight are visible around the Moon's edge. These spots are caused by sunlight passing through valleys on the Moon's surface. The effect is named after Francis Baily, an English astronomer who first described it in 1836. Continue using eye protection during this phase of a total solar eclipse.

Chromosphere: The layer of the Sun's atmosphere located above the photosphere and below the corona. The chromosphere is visible as a reddish ring during total solar eclipses.

Corona: Outermost layer of the Sun's atmosphere. Visible during a total solar eclipse as a halo of hot, ionized gas, called plasma.

Diamond Ring: A phenomenon observed during a solar eclipse when a single, bright spot of sunlight (the "diamond") appears next to the dark silhouette of the Moon, creating the appearance of a ring. Continue using eye protection during this phase of a total solar eclipse.

Ecliptic: The apparent path of the Sun across the sky over the course of a year, which is also the plane of Earth's orbit around the Sun.

First Contact (C1): When the Moon's disk first begins to cover the Sun's disk during a solar eclipse.

Fourth Contact (C4): When the Moon's disk completely moves away from the Sun's disk, signaling the end of a solar eclipse.

Hybrid Eclipse: A rare type of solar eclipse that appears as a total eclipse along one portion of its path and an annular eclipse along another portion. This occurs when the Earth's curvature causes the Moon's apparent size to vary.

Limb: The edge of the visible disk of a celestial body, such as the Sun or the Moon.

Lunar Distance: The average distance between the Earth and the Moon, approximately 384,400 kilometers (238,855 miles).

Lunar Perigee: The point in the Moon's orbit when it is closest to Earth. When a solar eclipse occurs, the moon is at perigee.

Lunar Umbra: The dark, central part of the Moon's shadow cast on Earth during a solar eclipse.

Magnitude: The fraction of the Sun's diameter covered by the Moon.

New Moon: The phase of the Moon when it is in conjunction with the Sun and not visible from Earth. A solar eclipse can only occur during a new Moon.

Nodes: The points where the Moon's orbit intersects the ecliptic. Solar eclipses occur when a new Moon is near one of these nodes.

Partial Eclipse: A solar eclipse in which the Moon's disk covers only a portion of the Sun's disk, leaving a crescent of sunlight visible.

Penumbra: The lighter, outer part of a shadow cast by a celestial object. In a solar eclipse, this is the area where only a partial eclipse is visible.

Photosphere: The visible surface of the Sun, from which light and heat are emitted.

Prominences: Large, bright structures in the Sun's corona, composed of ionized gas. They are visible as red, glowing arches during a total solar eclipse.

Saros Cycle: A period of approximately 18 years, 11 days, and 8 hours, after which the relative positions of the Sun, Earth, and Moon repeat, resulting in a similar pattern of eclipses.

Second Contact (C2): When the Moon's disk completely covers the Sun's disk, marking the beginning of totality during a total solar eclipse.

Shadow Bands: Faint, wavy patterns of alternating light and dark that can be seen on the ground just before and after totality during a solar eclipse. They are caused by the Earth's atmosphere refracting sunlight.

Solar Eclipse: A celestial event in which the Moon passes between the Earth and the Sun, casting a shadow on Earth and obscuring the Sun's light either partially or completely.

Solar Filter: A specially designed filter or material used to safely view the Sun during a solar eclipse, blocking harmful ultraviolet and infrared radiation.

Sunspot: A temporary, dark spot on the Sun's photosphere, caused by a concentration of magnetic fields. Sunspots are visible as dark, irregularly-shaped areas during a solar eclipse.

Third Contact (C3): When the Moon's disk begins to move away from the Sun's disk, marking the end of totality during a total solar eclipse.

Total Solar Eclipse: A solar eclipse in which the Moon's disk completely covers the Sun's disk, causing the sky to darken and revealing the Sun's corona.

Totality: The period during a total solar eclipse when the Moon's disk completely covers the Sun's disk, and the observer is within the path of the lunar umbra. You will be unable to see any of the suns rays through your eye protection during totality. It is safe to remove your eye protection during totality.

Made in the USA
Coppell, TX
08 November 2023

23963907R00105